Date Due

BRODART.　　　Cat. No. 23 233　　　Printed in U.S.A.

ELECTRONIC HEARTH

ELECTRONIC HEARTH

Creating an
American Television Culture

CECELIA TICHI

New York   Oxford
OXFORD UNIVERSITY PRESS
·1991

Oxford University Press

Oxford   New York   Toronto
Delhi   Bombay   Calcutta   Madras   Karachi
Petaling Jaya   Singapore   Hong Kong   Tokyo
Nairobi   Dar es Salaam   Cape Town
Melbourne   Auckland

and associated companies in
Berlin   Ibadan

Copyright © 1991 by Cecelia Tichi

Published by Oxford University Press, Inc.,
200 Madison Avenue, New York, New York 10016

Oxford is a registered trademark of Oxford University Press

Library of Congress Cataloging-in-Publication Data
Tichi, Cecelia, 1942–
Electronic hearth : creating an American television culture / Cecelia Tichi.
p. cm.   ISBN 0–19–506549–2 (alk. paper)
1. Television and family—United States.
2. Television and children—United States.
3. Television broadcasting—Social aspects—United States.
4. Television—Psychological aspects.
I. Title.
HQ520.T53   1991
302.23'45—dc20   91–10640

9 8 7 6 5 4 3 2 1

Printed in the United States of America
on acid-free paper

To Bill, Claire, Julia

For most people there are only two places in the world.
Where they live and their TV set.
                    Don DeLillo, *White Noise*, 1985

I was born in a house with the television always on.
          The Talking Heads, "Love for Sale," 1986

Home by the tube is where America takes place these days.
                Russell Baker, "Out of the Suitcase," 1989

Acknowledgments

Scholarship is collaborative, and this study is indebted to many who have taken time to read typescript, to offer advice and encouragement, and to challenge the positions argued. Among my colleagues and friends at Vanderbilt University, Joseph Urgo (now at Bryant College) read and advised me on drafts of most chapters, as did Jay Clayton, whose suggestions on subject positions proved invaluable, while Vivien Fryd gave advice on reading certain visual images. Graduate students at Vanderbilt University generously alerted me to relevant texts, and I thank Lance Bacon, Linda Barnes, Ken Cooper, Alan Lewis, Adam Meyer, and Lee Moore. In addition to their scouting, Chris Metress helped immeasurably as a research assistant in two succeeding summers. Our lunches at the Elliston Soda Shop have been sustaining in more ways than one.

I am thankful to two editors, Margaret Ann Roth of *The Boston Review* and Gordon Hutner of *Amercian Literary History*, both of whom published portions of this study. James Schwoch of the Department of Radio, Film and Television at Northwestern University gave the entire manuscript a scrupulous critical reading, and his comments aided immeasurably in challenging univocal presumptions. Portions of this work were presented at university talks and conferences, and I am thankful to respondents at the 1989 Wyoming Conference on English, Laramie, Wyoming, to members of the Institute for Liberal Studies at Emory University, to the 1990 Colloquium in English at Murray State University, Murray, Kentucky, to the Center for Interdisciplinary Research in the Arts, Northwestern University, especially to Rick Rodrick, Russell Reising, and Kathleen Woodward. Along the way, individuals suggested useful texts that might otherwise have escaped my attention, and I thank Judith Fryer, Rick Goodale-Sussen, and Gerald Graff. Amy Lang insisted that I recognize the role of alienation in individual male TV viewing, for which I thank her. And I thank Joan Levine for permission to use her phrase, "children in disguise."

The University Research Council of Vanderbilt University generously awarded funds for research assistance and photographs, and the staff at the Alexander Heard Library were most cooperative, as was the staff of the Vanderbilt University Television News Archive. The Learning Resource Center at Vanderbilt University photographed hundreds of images, and I especially thank Jamie Adams and Michael McCrickard. The staff at Mills Bookstore, Nashville, Tennessee, especially Gary Parker, alerted me to new titles pertinent to this study. Whittle Communications kindly expedited a copy of their publication, *Life After Television*, at the eleventh hour.

Within the habitat, a book-in-progress becomes atmospheric, a part of the household inhalations. Beyond the familial encouragement and forebearance, I appreciate William Tichi's assistance in navigating the formidable social science literature on children and television. Our daughters, Claire and Julia, became informants and research assistants on TV in rock music and videos. And the whole process has been made pleasurable by the attention of Rachel Toor, my editor at Oxford University Press. Her advice and careful reading of the manuscript have been first-rate.

Contents

ELECTRONIC HEARTH

Television Environment—A Preface

> *TV is environmental and imperceptible, like all environments.*
> Marshall McLuhan, *Understanding Media:*
> *The Extensions of Man*, 1964

> *People are now born into the symbolic environment of television and*
> *live with its repetitive lessons throughout life. . . . Living with*
> *television means growing up in a symbolic environment.*
> George Gerbner, *American Media and Mass Culture*, 1987

> Mass media grew until it too became a kind of environment.
> Jerry Mander, *Four Arguments for the*
> *Elimination of Television*, 1977

Television as an environment? An environment? The notion might sound bizarre.

Terms like *box* and *tube* have become standard slang of the electronic age—but "environment" is no television synonym, not even after forty years of television saturation in the United States. And no wonder, for the term both indicates a total surrounding and yet, in reference to television, leaves one at a loss to define or describe it. Those slang metaphors are so much easier to deal with, reinforcing the idea of television as an object, a thing occupying finite space, usually in one's habitat. To call television a box or a tube, for instance, is to think of geometric solids, cones, or cubes of glass, metal, wood, and plastic relegated to a minor place in one's personal or cultural space. But "environment" makes television the space itself, an encompassing surrounding. As a spatial configuration, television-as-environment is as bold as it is baffling, provocative but lacking ready reference and thus apt to leave one mentally as blank as the turned-off screen itself.

Marshall McLuhan, as we notice, assured us in the early 1960s that this is as it should be. In his view, environments evade perception. They are transparently atmospheric, he argues, their content being the material culture of the age that precedes them, the very age they displace (ix). McLuhan's assurance of the imperceptibility of the TV environment, moreover, absolves us all of the awesome responsibilities incumbent in the very term, *environment*. For the term obligates one to the godlike acts of naming and of visualization, acts of interpretation. The environment cannot speak for itself but must be spoken for and about. As it changes, it requires a historical response demarcating change and its motive forces. And in the act of interpretation, the speaker becomes interactive with that environment, a part of it, even interventionist.

All of our more familiar environments—rural, urban, wilderness—elicit these kinds of interpretive acts as a matter of course. All, that is, are mediated. And although it may seem natural to respond, say, to a geophysical environment like the Grand Canyon with awe and reverence, in actuality at least a century of interpretive public discourse from explorers' accounts to travel brochures has conditioned visitors to expect to feel—and therefore to believe that they do indeed feel—reverential awe instead of, say, dismay or revulsion. From the forest wilderness to the TV screen, every environment is mediated; the TV environment, like the others, has been shaped by a multitude of interpretive texts. Yet possibly because the once so powerful theorist of media, McLuhan, decreed the electronic environment to be invisible, many subsequent critics and analysts have treated it so. Others, grouping TV with various technologies, have opposed a malevolent technological world to an ideal natural one, dismissing television as artificial, second hand, evil—implicitly antienvironmental (Mander, Bradbury).

Electronic Hearth argues that the TV environment can and must be made visible. Set aside the premise on imperceptibility, and McLuhan's assertion on the TV environment remains nonetheless in place, reinforced some twenty years later in Gerbner's statement that television is a symbolic environment into which one is now born and raised. The interpretive obligation to bring that environment into high visibility is thus more urgent than ever. We must be enabled to see it and ourselves within it. And the texts that bring the television environment into consciousness are available. In the past four decades, they have been steadily accumulating in voluminous public discourse that crosses lines of genre, of medium, of social strata and gender, and ranging from novels to humor cartoons. These texts have delineated the television surrounding, including its social, moral, and linguistic dimensions. Despite the abundance of these texts, the TV environment has been largely ignored. Written, it has gone unread, the pertinent texts left in an archival state, like records and documents in storage. Examined here, they reveal that the TV environment has been synonymous with the social construction of

New Yorker cover by Stan Hunt showing the TV environment. Courtesy *The New Yorker*. Reprinted by permission.

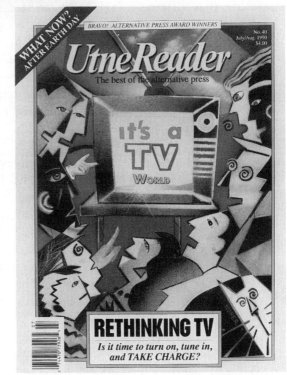

Utne Reader cover whose "It's a TV World" emphasis indicates the TV environment. Art by Peter Kuper. Reprinted by permission.

television, that we have environmental television traditions we have failed to notice or understand.

Electronic Hearth, subtitled *Creating an American Television Culture,* examines the TV environment according to the terms by which a major technology has been assimilated nationally over the past forty years. It documents the ways in which the public, constructed as a white middle-class American society, was prepared to accept a major new electronic technology of mass communication. This discussion acknowledges that the TV environment has emerged by the discourses of social critics, educators, the corporate world, advertisers, broadcasters, print journalists, and others. Their texts have sometimes been complementary, at other points radically competitive in their bids for assent to the version of reality they would purvey.

This is not a study of mass media per se, or an inquiry into the triad of radio–film–television. It is not a history of broadcast television or of politics in the electronic age, subjects that have been discussed in full elsewhere. *Electronic Hearth,* instead, draws on some forty years of advertisements, cartoon humor, art, journalism, memoir, and fiction to show the bases on which Americans have been led to understand the social changes—and the cultural continuities—of television.

Of course, television has been studied from numerous perspectives. One can configure it from within the industry, as the historian of broadcasting, Erik Barnouw, does, according to the human life cycle from toddler through elder. Or one can approach television history through the development of certain kinds of programming, as Edwin Diamond has done in his studies of TV news, and one can study the ways in which groups have manipulated television to further their own purposes. The sociologist Todd Gitlin has written extensively on this issue. Numerous other studies have presented the history of television from political, economic, feminist, or regulatory–legal–corporate perspectives.

Yet, these existing studies leave untouched the establishment of "the symbolic environment of television" (Gerbner and Gross, 443). In fact, the sources that best reveal that environment have not as yet been systematically examined. *Electronic Hearth* draws from forty years of public discourse on television, including texts in such mainstream American periodicals as *Life, The Saturday Evening Post, Playboy,* and *The New Yorker.* In these and other publications, journalism, humor cartoons, and advertisements for television receivers and other industry promotions reveal social attitudes that have formed the basis of the TV environment, from the receiver to the image.

The acculturation of television proceeds in and from such public discourse, in the kinds of texts utilized throughout this discussion, from fiction to advertisements. These texts are instruments constantly at work interpretively to position television in the cultural life of the public. In fact, one analogy drawn between television and the automobile is apt because, as is well known, the automobile is not only a transportation medium but layered with meanings on erotics, power, personal freedom, social status, and so on (Wallace, "Fictional Futures and the Conspicuously Young"). So too has television been imbued with sociocultural meanings both ascribed and assigned to it, and by the very kinds of texts just named, from cartoons to short stories. These texts continuously mediate between the receiver and the on-screen worlds on the one hand, and the individual and group of "viewers" on the other. They speak about, and on behalf of, the object, television, and they represent and enact the cognitive experience of the viewer. As interpretive texts, they both shape consciousness and in turn reflect that shaping. Altogether, they disclose the social construction of television, the terms on which it has entered and taken its place—rather, its places—in the culture from the 1940s to the present. In sum, they constitute the TV environment.

It cannot be dissociated from other, more familiar environments. This discussion argues that no matter how strikingly new a technology may be, once introduced into society it becomes deeply enmeshed in long-term cultural traditions and conflicts. *Electronic Hearth* shows, for instance, the deep involvement of television in national values including American individualism, domesticity and patriotism. And it argues (again, by ads and cartoons, fiction and journalism) that television has been configured in the discourse of Cold War ideology. It probes related issues on the problematic relation between action, work and leisure in post-World War II American culture, and it addresses TV-era discourse on sexual politics and on the war waged against television by the culture of print.

Necessarily, this discussion takes cognizance of historical processes. One must address certain sociocultural changes that result from the continuous technological progress in television (e.g., color, remote control, cable systems, larger screen size), which have modified viewing practices and even cognitive processes. In addition, it is tremendously important to recognize that the TV age coincides with the era of the post-Cold War, extending from the late 1940s into the late 1980s. Because the cultural assimilation of a technology is also the story of its social construction, it was inevitable that Cold War issues should become television issues. Certain patterns have persisted, such as anxiety about invasion, about security, about the American home and family, about work and leisure. Some of these issues have been modified from the 1950s onward, as, for instance, the phenomenon of being televised gradually has taken on an importance of its own. Other patterns, however, have stayed remarkably consistent, for instance anxiety about television being at once an addiction and simultaneously the contemporary family hearth.

Any study undertaken at this historical moment must also confront the division between two TV generations. The first of these consciously experienced the introduction—even incursion or intrusion—of television into their lives and habitats and, therefore, understood it to be discontinuous, while the succeeding generation, their children, never knew life without the small screen and have experienced television as integral and natural. Any visitor to the Smithsonian Institution's 1989 exhibition, "American Television: From the Fair to the Family 1939–1989," could grasp this generational division. To stand at a railing before a display of early TV receivers was to hear older people recite in considerable detail—almost compulsively—their recollections of the first television broadcast they saw, the first TV set to be brought into their habitat. At length they would describe the cabinetry, their first programs, the recollected remarks of family members or friends in the immediate vicinity. The writer Bobbie Ann Mason (b. 1940) has recaptured this kind of moment in a story, "Detroit Skyline, 1949," in which a western Kentucky farm girl sees television for the first time when visiting an aunt and uncle:

> "It's one of those sets you can look at in normal light and not go blind," my aunt said, to reassure us. "It's called Daylight TV."
>
> "Wait till you see Howdy Doody," said Uncle Boone.
>
> The picture on the television set was not clear. The reception required some imagination, and the pictures frequently dissolved, but I could see Gorgeous George moving across the screen, his curls bouncing. I could see him catch hold of his opponent and wrestle him to the floor, holding him so tight I thought he would choke. (*Shiloh*, 39)

Such personal narratives as this, recited at the Smithsonian in 1989, were obviously compelling and nostalgic to the tellers and to their cohorts. But an observer listening to these narratives could not help noticing the responses of young listeners along for the museum trip. Standing with the older family members, they seemed distinctly unimpressed, enduring the TV stories in apparent boredom. Their position might best be summarized by the young writer, David Foster Wallace, who observes that "the American generation born after, say, 1955 is the first for whom television is something to be *lived with*, not just looked at . . . we, unlike any elders, have no memory of a world without such electronic definition. It's built in" ("Fictional Futures," 38–39). Built in—and thus unremarkable.

But not unimportant. *Electronic Hearth* exploits recent fiction and visual images now available from a generation of writers and artists who have grown up in the TV environment and are now reflecting—and reflecting upon—its forms, traits, values, and in turn enacting them in their own work. Because it is undertaken in the late 1980s and early 1990s, this book is able to exploit a sizable body of TV-related source material that has only come available very recently—memoirs, short stories, novels, comics, and videos of a younger generation of writers and artists with lifelong biographical immersion in the world of television.

Thus, the texts on which *Electronic Hearth* is based include the insiders' version of the TV environment, the representations of the cognitive and perceptual acts that define a generation, so to speak, "born in a house with the television always on," as the Talking Heads lyric says. Back in 1930, the American cultural critic, Matthew Josephson, half-facetiously suggested that American artists and intellectuals would one day join forces with the mass media to become "centurians of soap . . . [and] dance before the television box [in order to] leaven this society" (307–8). Dancing aside, Josephson was correct in his assumption that popular and elite cultural forms would merge in the television era, and *Electronic Hearth* works from the assumption that the culture of television embraces numerous and diverse texts in an era when the boundaries between different kinds of previously stratified texts have become porous.

The reader of *Electronic Hearth* will find a theoretic mixture. This study has recourse to numerous theorists, from those expounding ideas on hyperreality, notably Jean Baudrillard and Umberto Eco, through Frankfurt School students such as David Antin, together with the cultural studies of Roland Barthes, especially his self-styled "mythologies" (*Mythologies, The Eiffel Tower*). It has been necessary, however, to resist univocal adherence to any single theoretical school because the territorial claims of each would preclude the insights enabled by others. To embrace the Frankfurt School position on the hegemony of state capitalism, for instance, is necessarily to see the television viewer as a figure essentially passive, held captive by the communications apparatus of the capitalist interests, at best a prisoner of false consciousness. Yet textual evidence shows decades of personal resistance to television, the very kind of resistance that Michel de Certeau discusses in terms of "the practice of everyday life." The resistant figure, however, has no legitimate place in Frankfurt School thought, which reduces him or her to a figure co-opted by corporate media interests. It seems most accurate to argue at this point that a viewer might at different times be a resistant viewer and then a corporate-capitalist subject, and at other points a hyperrealist. As Mimi White writes, "Television itself is a mass, industrial medium, involving a variety of texts, produced by many different groups (and individuals), aimed at a broad and heterogeneous set of audiences. It thus becomes difficult to talk about a single set of beliefs or ideas that are carried by television in any simple or immediate sense" (136). At a time when univocal positions of every kind are under challenge, when discourses are seen to serve varying interests and the recipients of messages from every quarter are thought to be comprised of individuals

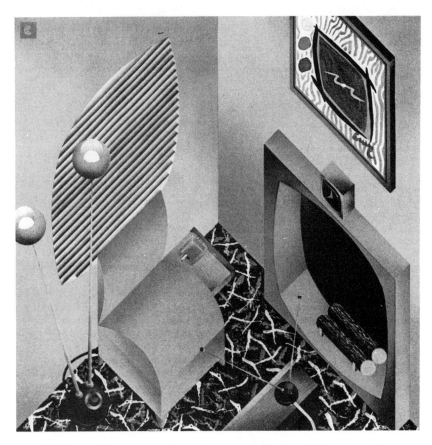

Atlantic Monthly magazine illustration in which every object from fireplace to window blinds refers to television.

defining themselves in myriad self-identifications, converging only intermittently in groups—at such a point the analyst can surely claim latitude in theoretics.

We remain at this time midway in the process of assimilating television into the culture, as the technology progresses and continuously modifies the viewers' relation to the on-screen world (e.g., by the personal camcorder–VCR, the imminent high-definition television, and interactive television undergoing trials in some cable markets). If a comprehensive grasp of the TV environment is conceptually premature, the moment is right for a midcourse analysis.

Introduction—Phasing In

Our Parents regard the set rather as the Flapper did the automobile: a curiosity turned treat turned seduction. For us, their children, TV's is as much a part of reality as Toyotas and gridlock. We quite literally cannot "imagine" life without it.
David Foster Wallace, "Fictional Futures and
the Conspicuously Young," 1988

In a few strokes, a young writer graphs the evolution of television in the United States. His parents' generation, the post-World War II counterparts of 1920s Flappers, confronted the exotic new TV as a novelty or curiosity, something like a gift or present that finally made them feel debauched. Their children, all the while, experienced it quite differently. The young, born into a world of ubiquitous TV sets, have understood television from the first as an unremarkable part of the natural order of things, "something to be *lived with*, not just looked at" (Wallace, "Fictional Futures," 38).

The young writer is talking about the development of the TV environment in the United States. To speak of the evolution of television from the exotic to the commonplace is to refer to the process of its sociocultural naturalization, the process by which a phenomenon, initially conspicuous and dislocative in its very newness, becomes over time sufficiently assimilated that people experience it as a part of the natural order of things.

But the naturalization of television, like that of many technologies, has occurred so gradually, so much on a continuum that the step-by-step demarcations blur, making it difficult to gauge how, when, and where the TV environment has come into being, and on what terms. To notice certain divergent viewpoints that bracket the TV age thus far is to see the scope of this naturalization. Social commentators' predictions on television in the late 1940s, for instance, sound like a time capsule of anxieties to be unearthed for scrutiny in the last decade of this century. Television, it was claimed, would totally destroy radio and movies, "end the art of conversation" and bring domestic life to a "standstill" and thus undermine "the American way of life" (Hornaday).

The very impulse to "correct" the 1949 statement is a measure of the thorough acculturation of television in the United States. The urge to bring the forecast up to date, or, conversely, to ridicule it as archaic springs from the same sense of the disparity between the projected TV era and the actual one— which a 1989 pundit presumes to be television-centric as he writes, "To stay in touch with America you had to stay home glued to the tube" (Baker, "Out of the Suitcase"). As these two kinds of positions indicate, over forty and more years the television environment has moved from a realm of futuristic fantasy to a presumption of material actuality. Still, the course of that movement has not as yet been charted, which is to say that the TV environment has neither been brought into visibility nor observed in its changes over time. It is high time to focus—to bring into consciousness—that process, which is one of acculturation of a major technology, one by which the TV environment takes shape.

Initiation

We begin in 1943, a wartime point by which Western military and civilian leaders knew that the Allies would ultimately triumph in the second world war of the century, and that the U.S. economy would reconvert to peacetime manufacturing. Looking ahead, the Allen B. DuMont Corporation initiated a remarkable series of advertisements that set the terms in which television would be presented to the postwar public, itself conceived as a consumer market. DuMont had a major stake in setting the terms by which broadcast television would be presented to this public, since the DuMont Laboratories had developed television-related technology, had the capacity to manufacture receivers for the postwar retail market, and would soon control a major network and thus be positioned to reap huge sums in advertising revenues. Within five years, Pittsburgh viewers would hear station breaks announced in a portentous male voice intoning in rhythmic cadence, "This is the DuMont Television Network." But, even in 1943, the company's ads typified corporate America's bid to shape the TV environment in ways that made the buyer of a receiver feel modern, more than that, utopian.

Television, said DuMont in 1943, would mean the peacetime application of technology currently militarily successful against German submarine torpedos and the Japanese navy. "Wartime Inventions with Peaceful Intentions," proclaimed DuMont of its oscilloscope, cathode tube, and cyclograph.

Conjoining these technologies from war to peacetime usage, DuMont made television into a promise of continuing patriotism and technical expertise. And the company familiarized the televisual process in wartime patriotic figures of electrons moving in "squads" under the direction of "sergeants," the electronic scanning likened to reading a book page, and therefore, not really exotic, much less bellicose. Yet in appealing also to the masculine premise of the American Western gunslinger, television was figured as a more powerful Deadeye Dick shooting his cathode ray bullets with "incredible rapidity." In the forthcoming peacetime, the ads implied, television would deploy the most sophisticated technology in the service of the highest national values, these configured as aggressively masculine (*Harper's Magazine*, 186 [May, 1943]: n.p.; 187 [June, Aug., Nov., 1943]: n.p.).

At the same time, DuMont moved to demystify the exotic new TV technology, using the Hollywood actress Paulette Goddard and a cartoon vaudevillian named Alec Electron who resembled Fred Astaire in top hat and morning coat, a guide through step-by-step panels on the relaying of the film star's image to the television screen (and thereby implying the imminent succession of Hollywood by television). Subsequent ads featured Alec Electron answering consumer questions on when television would be ready, when receivers would go on sale, what screen size and quality of picture could be expected, and so on (*Harper's Magazine*, 188 [Dec., Jan., 1943; April, 1944]: n.p.).

All the TV metaphors to be exploited in the following decades are present in this ad series. Television is "The biggest window in the world," a "looking glass" through which the viewer becomes "a modern Alice." It is "the greatest show on earth" and located in Shakespearean metaphor—"'All the World's a stage' . . . with Television." It is a "gift," "the answer to man's ageless yearning for eyes and ears to pierce the barrier of distance." It is both "magic carpet ride" and "university." "Television cameras will be your eyes" and will "topple the tower of Babel." And celebrities were shown in on-screen photos endorsing the new medium—the singer Kate Smith, the comedians Fred Allen and Jimmy Durante, the "incendiary blonde" movie star Betty Hutton promising to "be practically in your lap."

Celebrity endorsements aside, these metaphors ring with utopian fervor. They promise the fulfillment of aspirations stated to be transcendent of history and of geography ("the answer to man's ageless yearning"). The corporation conveyed the message that television and the American future of material and spiritual progress were synonymous, and this rhetoric, in part the legacy of Protestant millennialism, was now applied to television as it had been used to evoke the utopian goals of electrification and radio, and would evoke them still in connection with plastics and nuclear energy (see Corn, Segal).

WHAT IS TELEVISION?

Alec Electron is back and this time he has his own over-simplified explanation of this question

Here I am again - Little Alec Electron

I'm the works in Television—and how!

Paulette Goddard* is being televised by a camera which has a vacuum tube with a flat surface of tiny particles of an unusual substance. Here we electrons are free. When light hits us, some of us absorb enough energy to jump.

When the light of Paulette Goddard's image falls on this surface, we jump with joy, and jump right off! It's really the light that makes us leave; and the stronger the light, on each point of the image, the more of us must jump.

Now, this tube has a squad of electrons who check up on us. They act as a Cathode Ray beam, and with it thirty times each second they scan the entire surface left to right, top to bottom, as you'd read the page of a book.

They send out messages telling just when each new "scanning" starts (scientists call it the synchronizing signal). As they scan the picture they also send information as to the exact number of us jumping from each point.

These messages, both starting signals and the number of electrons leaving each point are sent to the television transmitter and pumped out as radio waves—instantly intercepted by your antenna and sent into your receiving set.

The radio wave orders are picked up by Grid Sergeants in the Cathode Ray Tube in your receiver. This Cathode Ray Tube is the heart of the television set, commercially developed by DuMont. As the Sergeants respond to the radio signal they maneuver us to

scan the fluorescent screen in perfect coordination with the electron beam in the camera tube. As directed, they send exactly the same number of us to each point of the screen in your set, as was on each point of the screen in the television camera. Thus making a picture.

That's how we put a picture in your television set almost at the instant it is "scanned" in the studio. Back in the station those M.P.'s send out 30 pictures a second. Of course, nobody's eyes work that fast, so the effect you get is of continuous moving pictures—or television.

* Soon to appear in Paramount's "Standing Room Only"

© 1943 Allen B. DuMont Labs., Inc.

Allen B. DuMont, the man who made television practical, is awaiting only peace to produce your home television receiver

Allen B. DuMont Laboratories, Inc., General Offices and Main Plant, 2 Main Avenue, Passaic, N. J. Television Broadcasting Studios and Station W2XWV, 515 Madison Avenue, New York, N. Y.

DuMont advertisement, 1943.

YOU'LL BE AN ARMCHAIR COLUMBUS!

You'll sail with television through vanishing horizons into exciting new worlds. You'll be an intimate of the great and near-great. You'll sit at speakers' tables at historic functions, down front at every sporting event, at all top-flight entertainment. News flashes will bring you eye-coverage of parades, fires and floods; of everything odd, unusual and wonderful, just as though you were on the spot. And far-sighted industry will show you previews of new products, new delights ahead.

All this — the world actually served to you on a silver screen — will be most enjoyably yours when you possess a DuMont Television-Radio Receiver. It was DuMont who gave really *clear* picture reception to television. It will be DuMont to whom you will turn in peace-time for the finest television receiving sets and the truest television reception ... the touchstone that will make you an armchair Columbus on ten-thousand-and-one thrilling voyages of discovery!

DuMONT *Precision Electronics and Television*

ALLEN B. DuMONT LABORATORIES, INC., GENERAL OFFICES AND PLANT, 2 MAIN AVENUE, PASSAIC, N. J.
TELEVISION STUDIOS AND STATION W2XWV, 515 MADISON AVENUE, NEW YORK 22, NEW YORK

DuMont advertisement, 1944.

The profusion of these metaphors, applied to television, becomes an exercise in the simultaneous expansion and containment of meaning. True, each figure from Shakespearean stage to looking glass floats in the public domain and is reassuring for that very reason. By association, Shakespeare imparts universal artistic greatness to television, while Lewis Carroll's Alice certifies its appropriateness for children. These images not only embrace the widest range of sociocultural associations and bond them to television, but work as well to blunt the extremes of meaning encoded in each metaphor. A technology linked with Shakespeare must not be allowed to seem effete or intellectual. A technology to be brought into household use must not be identified too closely with ballistics, whether torpedos or Wild West sharp-shooters' bullets. As pure entertainment ("the greatest show on earth") endorsed by radio and film stars, television must not, however, present itself as trivial. And American men must not be identified too closely with Alice in Wonderland and a fairy tale world of magic carpets. Thus, television must occupy imaginative space at the borne of all of these metaphors, participating in them and yet not fully committed to any of them. The ad texts cultivate an ambiguity meant to reassure all respondents. Simultaneously the images are deployed to check each other—for instance, the fairy tale and its amusement ("magic carpet") against institutional education ("university"). And the domi-nant masculinity (Deadeye Dick) allowed the feminine as it was ratified by men (the Astaire-like Alec Electron) (*Harper's Magazine*, 188 [May, July, Aug., Dec., 1944]: n.p.; 190 [Feb., March, June, 1945]: n.p.; 191 [July, Aug., Oct., 1945]: n.p.).

DuMont projected peacetime and television together in a pastoral vision that originated in those wartime ads. To a public exposed weekly in movie theatres to newsreel footage of bombed cities from London to Dresden, DuMont promised "the [peacetime] country home you've always longed for . . . in an era in which America will be linked through television with the most remote peoples of the earth." The corporation's advertising twice featured photographs of Winston Churchill, together with a quotation in which the wartime British Prime Minister cited "television" as one medium to make "life in the country and on the land . . . compete in attractiveness with life in the great cities."

In the postwar years, as is well known, the country-home ideal would realize itself in the form of the American suburb in which the consumerist "showcase" home included garage doors that opened by remote control, wall-to-wall carpeting of nylon, air conditioning, a plethora of small electrical appliances like the Mixmaster—and in the era of nuptial "togetherness," "a smart new marriage of television and decorating," as one advertisement put it. If the so-called marriage reinforced the marital state of the suburbanites, it is useful to

The enthroned white man in TV recliner chair.

notice the presumption that the provider or bread winner, the husband–father, controlled the purchase of so-called major appliances including the TV set. His earnings would pay for it. His wife might be the arbiter of taste, choosing the furniture styles, but the man of the house had to be persuaded to part with the money—especially for such a major purchase. If the advertising layouts emphasized the woman's sphere, in *Ways of Seeing*, John Berger reminds us that the pictorial representation of women and household objects is a way in which prosperous men can display their possessions. They themselves need not necessarily appear in the picture, for it is assumed that the man, in gazing at the scene of his family amid his household property, looks upon what he possesses. In the TV advertising texts, we can infer the same relation. The illustrations showing comfortable suburban interiors pay tribute to the man as good provider and master of family and property. In fact, a variation of this message was to be found in texts promoting the furniture specifically developed for the TV viewer, the recliner chair, which customarily was shown with a man settled comfortably after work in the reclining position, gazing undisturbed toward the TV screen, the household monarch enthroned ("Show House" advertisement, *Saturday Evening Post*, 227 [April 23, 1955]: 130–131).

Carapace

To vocalize the names by which television receivers were marketed from the late 1940s through the 1970s is to understand that maximal profits went hand in glove with upscale positioning of the television set. Manufacturers and their advertising cohorts made the TV receiver the objectification of consumerist prestige and social status. Assigning the receiver a name and housing it in cabinetry reminiscent of furniture styles familiar to connoisseurs of the decorative arts conveyed a distinct message about how consumers ought to regard their social status, wealth, and aspirations.

A review of the names of the receiver carapace shows the sociocultural terms on which individuals and families were led to accept—even to desire—the receiver itself. Representative model names include: the Admiral Corporation's The Jefferson and The Nording, The Nassau, The Honduras, The Continental, The Houston; RCA's The Yorkshire, The Glenview, The Abington, The Cherbourg, The Hillcrest, The Blithe, The Masterson, The Baylor, The Princeton, The Vermeer, The Kenniston; Philco's The Rockland, the Steffens, and the Miss America Series; DuMont's The Lancaster, The Richfield Modern, The Hanover, The Revere, The Plymouth, The Devonshire; Magnavox's The Aegean Classic, The Shoreham, The Wedgewood, The Constellation II. No receiver lacked its designative article, "The," indicating uniqueness and status as art object, as in *The Mona Lisa* or *The Elgin Marbles*. (Portable receivers typically had less imposing names; e.g., Zenith's Trend-setter, RCA's Sportabouts, Admiral's Playmates. Taking up far less physical space than the console and floor models, they were presented to the public in very different terms.)

As for the floor models and consoles, the nomenclature is strikingly Anglo-British (The Devonshire) and Continental-European (The Cherbourg), and American Colonial and Revolutionary (The Plymouth). All these names invited associations with high social status and cosmopolitanism, this last especially from the early 1960s, when the Kennedy family were shown on TV and in magazine layouts internationally sojourning, and when airlines with transcontinental routes heavily promoted European, Caribbean, and Mediterranean travel.

The point is that television, during wartime connected with precision radar and explosives, enters the postwar American household, not as technology but as elegant furniture suitable for the "country home." Why?—because the middle class American woman needed persuading that a large box approximately the size of a washing machine, containing an exotic-sounding technology called the cathode ray tube, was neither intrusive nor invasive in "her" most respectable living space, that on the contrary it was enhancing of that space and ought to be accorded a prominent position as focal object. The DuMont Westminster, for instance, marketed in 1946, was 38 inches high, 64 inches wide, 26 inches deep. A 1952 magazine article on "Television: Giant in the Living Room," and a 1955 newsmagazine headline on "Goliath with tubes" precisely caught this implication of something huge and massive, the very image that the industry sought to avoid and counteract (Willingham, 114; *U.S. News and World Report*, Sept. 2, 1955, p. 37).

Advertisers' texts thus emphasized TV set design said to be in accordance with women's judgment. "Because it is designed from a woman's point of view," said a Stromberg-Carlson ad copy text of 1951, "this superb instrument brings distinctive good looks as well as matchless performance into your home." Significantly, the "good looks," not of the mundane "receiver" or "set" but the "instrument," were rooted in historical periods, in Regency or Queen Anne or French Empire or in the American colonial eighteenth century. The newest telecommunications technology was entering the American home both on the basis of its anchorage in the past and its simultaneous claim to its ahistorical status as unassailable classic (*Life*, 31 [19] [Nov. 5, 1951]: 110).

True, the atomic-age TV set, "futurized" as the Raytheon Corporation proclaimed, appealed to the idea of technological advance, and some promotional TV images featured the exposed receiver, "TV's greatest power plant" with tubes and wiring on view, doubtless intended to appeal to men accustomed throughout the century to similarly formatted advertisements for automotive products. Yet, these kinds of images were few. A historian of design remarks, "Unlike the radio and other sound equipment, which could be located inconspicuously, the TV had to be located centrally at a distance and a height that was convenient for viewing. . . . The biggest problem in designing a TV was what to do with the depth of the tube" (Pulos, 299, 301). In terms of consumer acceptability, the tube, technologically the *sine qua non* of television, was a potential liability, large and protuberant. Home decoration magazines, serving the interests of furniture as well as TV manufacturers, assured women readers, "You don't have to pull your house out of shape to make room for a television set . . . your television set can be inconspicuous when it's off . . . [but] if you want television to dominate the room, place set against a simple background" (*House Beautiful*, 91 [Aug., 1949]: 66–69).

Technological emphasis on the era of TV as the Atomic Age.

Such texts assured the consumer of the ability to control the space in which the television would be positioned, in effect to control the very size of the receiver. The consumer could "put television in its proper place," the mastery assigned to the consumer.

But the new television lexicon could sound formidably technical, and the new, alien terminology also had to be domesticated to make television acceptable in the habitat, especially to women. In 1950, *House Beautiful* magazine offered its women readers a dictionary of "typical TV lingo," terms like electronics, sending tube, chassis, fluorescent screen, electron gun, radio spectrum (92 [Aug., 1950]: 110–11). This lexicon was useful, readers were assured, in order that you, the householder, could understand the language of "your serviceman," a figure implicitly in "your" household staff, like the chauffeur. Such texts assumed that it was women's self-interest in social status, not interest in masculine technical knowledge, that would prompt her mastery of such terms as "electron gun" or "kinescope tube." The TV manufacturers and advertisers, together with the periodicals that they supported, put technics into the service of social hierarchy. And in the women's sphere, the problems of intrusive masculine technology were also solved by disguising the TV—or rather, transcending it by changing the terms of its presentation from machine to aesthetic object. Promotional texts featuring exposed electronics always carried inset photographs of handsome cabinetry in which the mechanism was hidden away, positioned in the habitat with draperies, carpeting, and the like, showing that the TV receiver bore no resemblance whatsoever to a machine. As Lynn Spigel observes, the installation of a TV receiver meant a "theatricalization . . . of domestic space" (12).

In part, because of sheer size, the receivers were customarily displayed amid the signifiers of cultured, civilized life of sophisticates, including old books understood by their uniform bindings to be "classics," and flowers or potted plants signifying controlled nature and a servant staff with a gardener, and objets d'art suggesting good taste and the leisure time in which to pursue collecting as an avocation. In 1955, DuMont suggested that the TV set was "both an entertainment center and part of a heart-winning group of flowers, books and cherished whatnots." "Handcrafted" or "handrubbed," said the ads, identifying TV cabinetry of mahogany, cherry, and other valuable woods and sometimes identifying by name the designer. "This superb furniture is also the world's most advanced color TV," proclaimed the Magnavox Corporation in 1968. DuMont said of its "Sherwood" model, "Only a few can own it." The idea of exclusivity, of handcraftmanship for the privileged elite suffuses these advertising texts, emphasizing that television was no mere household appliance, much less a mass-produced machine, but an objectification of the good taste and commitment to affluence of the modern American family. And

Illustration for Sears Silvertone television, 1966.

those consumers purchasing the less ornate and less costly consoles were expected to understand that the highest priced, largest receivers set the standard for television equipment, that essentially they defined television in American culture (*Life*, 63 [21] [Nov. 24, 1967]: 65; 65 [20] [Nov. 15, 1968]: 79; *Saturday Evening Post*, 227 [April 23, 1955]: 131).

It is worth looking closely at one ad text of this type, a 1966 advertisement for Sears color television, featuring a couple, "The Peter Duchins," he identified as "a famous pianist and society bandleader." The Duchins, as readers of the copy text learn, are shown photographed in the library of their New York City brownstone townhouse, and one immediately sees the wood-paneled walls with numerous framed pictures, *objets d'art*, books (including oversize art books), and fresh flowers. The text tells us the mantelpiece urns are eighteenth-century Chinese, the wood-carved idol Indian, and the pre-Columbian statue Mexican. Emphatically, this is a cosmopolitan couple of discerning tastes, and Cheray Duchin is quoted as saying that TV cabinetry must bear the Duchins' exacting aesthetic scrutiny ("We've collected some marvelous pieces of furniture during Peter's tours. I think our Sears Silvertone cabinet fits in beautifully. It's the French elegance model."). Her husband sits comfortably on the sofa by the Oriental rug gazing toward the television just as, on-screen, in an image identified in small script as "simulated," two figures on horseback pass by in a country scene echoed in a landscape oil painting over the fireplace. The hearth and television cross reference each other, and the Sears TV set gathers to it all the associations of the Duchins and their objects. The TV set, as furniture, means internationalism, privileged social status, irreproachable taste. All these attributes are understood to be transferable, via the television receiver, to those who acquire the Sears TV.

Television furniture, then, with all its sociocultural meanings, served to naturalize the cathode tube. Only in the late 1950s when manufacturers could feel assured that the newer portable televisions were "second sets" and, therefore, "did not have to pretend to be a piece of furniture," did advertisers dare to flout their own carefully constructed associations (Pulos, 301). In 1972, the Sony Corporation suggested that buyers of its portables could keep them atop the old furniture consoles ("So what should you do? . . . You should use the console. . . . It's a beautiful piece of furniture. Keep your Sony on it"). The portable design in its dematerialized state was broken free of the idea of furniture. At that point the television receiver, entrenched in the habitat, smuggled into it in the guise of furniture to facilitate technological change, was permitted to escape the huge carapace in which it had first traveled into the American habitat.

The Custom of the Country

It may not be possible to know whether television, newly come to the habitat, was as dislocative as certain texts suggest, but indications are that for the white middle class, existing codes of conduct did not readily accommodate the new technology. Instead, television was felt to necessitate new behavioral rules. "Most TV families eat with their plates balanced on one knee and the coffee safely out of the way under the chair. This can be verified by examining the condition of the living room rug," wrote one humorist who in 1951 chronicled the social dislocation of his family when "the monster" took its place in their living quarters (Ritts, 95). (The narrator, his career spent in radio, resisted the eulogistic nomenclature of television furniture models and so dubbed his receiver "the monster.") "Having a television set," he continued, "one automatically becomes the proprietor of a free theatre, free snack bar and public lounge," adding that "being alone was something of a novelty" (96).

As these lines indicate, television initially disrupted established manners, etiquette, behavioral norms of white middle class America. Essentially, private and public spaces were abruptly conflated and positioned in mutual contradiction. The private home suddenly acquired traits of public places— theatre, snack bar, lounge, stadium. Of course, that was the promise of the ads—"television will transform your easy chair into a choice orchestra seat at the theatre . . . into the finest box on the third base line when the empires shout, 'Play Ball!'" Anxieties arose, however, precisely because television did fulfill that promise, at least in significant part. The advertising texts did not prepare individuals for the resultant confusion. For customs and mores were disrupted when television reconfigured its surrounding space in new, ambiguous spatial, and therefore social, terms (DuMont advertisement, *Harper's Magazine*, 189 [Oct., 1944]: n.p.).

Television-centered food becomes one locus of the new ambiguity. A woman was encouraged by advertising texts to think of herself as the lady of the house seated for reposeful TV evenings of ballet and theatre. If guests were present, she was also a hostess. In 1947, the *New Yorker* magazine published a week-long diary of a viewer–reporter invited into a private home for several viewings of broadcasts from the three operating New York City stations, whose programs ranged from Brooklyn Dodgers games to the Kraft Television Theatre. In print, the *New Yorker* writer compliments the woman of the household as "one of the most bountiful hostesses I have ever met." She "passed enormous helpings of ice cream and cake. Peanuts, popcorn, pretzels, and highballs had been circulating all evening, and on a coffee table was a prettily arranged bowl of fresh fruit" ("Diary of a Viewer," 51).

It might have been interesting to hear this "bountiful" hostess's side of things. Several texts suggest that with the installation of television, the conflation of private and public space threatened to turn this lady–hostess into the cook and waitress preparing and serving food to others indifferent to her, absorbed as they were in programs she could but glimpse en route to and from the kitchen. These others she served were not only her invited friends or her family to whom she would feel obligated as wife and mother to prepare meals or snacks. "Others," instead, were a quasi-public aggregation identified as "neighbors." "The beams of our living room groaned with the weight of half the neighborhood population . . . one couple I never did meet. . . . How they got in I'll never know" (Ritts, 98). On this point the aptly titled *The TV Jeebies* names "the Moocher" as one of the TV boors: "this is the fellow who picks his seat next to the tea table where the refreshments reside, and never leaves it" (105).

As the rules of hospitality clashed with the norms of public behavior, the promotional press attempted to defuse conflict. In 1960, the editors of *TV Guide* published a miscellany of material previously published in the weekly magazine, including a recipe section entitled "TV Snack Time . . . For a

"Did you hear that fat guy, the one that sat in the green chair, ask me if I lived in the neighborhood?"

THE SATURDAY EVENING POST

Cartoon from *Saturday Evening Post*, 1951. Reprinted courtesy of © Curtis Publishing Co.

Harried Hostess." "Today's hostess need not be harried," the woman reader was assured in notes prefacing recipes for "the kind [of food] you may eat with your fingers." "Prepare snacks at leisure to serve in haste when expected or unexpected visitors drop in," wrote the food editor in the employ of the magazine (*TV Guide Roundup*, 194ff). The "moocher" is disallowed in this text, in which all entrants into the household are "visitors," meaning guests, even if they "drop in." The hostess is sufficiently "at leisure" to prepare TV food in advance of actual viewing time, and it is assumed in this promotional text that she will wish to do so and obligingly serve it "in haste" as directed. Should she transfer food preparation to other, commercial hands, the responsibility nonetheless remains hers, and in that same year, 1960, the Campbell Soup Company accordingly proclaimed that their frozen Swanson TV Brand Dinner contained only "choice slices of tender turkey, carved *exclusively* from meaty breasts and thighs." A novelist (here, John Updike) can register the critical judgment on the TV dinner "Salisbury steak that tastes of preservative," but the advertiser, once again, puts the idea of the exclusive into play in the relation between viewers and their TV receivers. It is a term of assurance of nutritional and social status doubtless intended to offset the visual image of the

Television as a theatrical evening.

The TV-era hostess in Magnavox advertisement, 1950.

Trust Swanson

Who else gives you only choice slices of tender turkey?

There's no comparison! A Swanson TV Brand Dinner gives you such choice slices of tender turkey, carved exclusively from meaty breasts and thighs. Swanson adds garden peas, milk and butter whipped potatoes, and real cornbread dressing with rich brown gravy. For good frozen dinners, trust Swanson!

Campbell Soup Co. advertisement for TV dinner.

frozen-dinner container as an institutional cafeteria tray. Campbell preferred one not to notice that an object identified with a lower-order public eating place was now proferred in the most intimate, private setting of the family (*Rabbit Redux*, 28, *Life*, 49 [23] [Dec. 5, 1960]: 47).

Although the *TV Guide* miscellany smoothed over social disjunctions, it also disclosed certain terms of disruption in the first decade of mass-marketed television. The lines of tension from the new technology were revealed in a self-help quiz on "your TV etiquette," in which the reader–viewer was invited to "test your social reactions to the following situation." Two of the quiz questions with the "experts'" answers ran as follows:

> PROBLEM 5: You have guests in your home and, either by design or special request, television has been turned on. When the program is over, do you:
>   (a) Turn off the set promptly?
>   (b) Let the next program come on and see what your guests will do?
>   (c) Check the listings, take a poll to see if they want to continue viewing, then let the majority rule?
>   (d) Bring out some snacks and hope for the best?
>
> Answers: PROBLEM 5: (a) courageous but not too gracious; (b) shiftless; (c) most desirable; (d) hopeless.
>
> PROBLEM 6: There's a special ninety-minute TV show scheduled. You're anticipating it so keenly that you decide to invite a dozen congenial friends for dinner and the program. However, half of the proposed guests politely indicate they prefer not to watch TV. Would you:
>   (a) Tell them to come anyway, then try to find some other diversion for them while the show is on?
>   (b) Tell them that though you'd like to see them you can't change your TV plans and ask them to go along with your arrangements?
>   (c) Forget the whole project?
>   (d) Invite the six who don't want to watch TV for another evening when television will not be an integral part of the occasion?
>
> *Answers:* PROBLEM 6: (a) ninety minutes is too long to split up your group; (b) you're likely to have some unhappy guests; (c) gives the impression you're miffed; (d) most advantageous and comfortable for everybody. (121–24)

The quiz, a primer on TV etiquette, is based on competing allegiances between television and oneself on one side, and the social group on the other. The suppositions of the other quiz questions reinforce this division: "The telephone rings. It's an invitation for dinner on the evening of your favorite show." What to do? Or, "You are watching television. Something you really don't want to miss. About halfway through, uninvited guests drop in." What to do?

The tension is repeatedly presented as conflict between the TV viewer entitled to his or her viewing, and the disruptions of human relationships. The quiz implies that, between them, the quiz reader and *TV Guide* editors understand that television itself is good company, that one has the right to watch it without interruption. Yet the quiz also acknowledges what reader and editors also know, that the prerogative to watch will be misunderstood by others as selfishness, as antisocial or undemocratic behavior. To be an adult viewer is to be repeatedly presented with versions of the same dilemma. To state that the self-and-TV are a self-contained twosome is out of the question, as is refusal to accommodate the social situation. One's right, in short, is not one's prerogative. Between the quiz text and its reader lies the understanding that the viewer/TV owner is vulnerable through no fault of his or her own, that the situation of the modern TV viewer risks flaring into conflict that others will perceive as selfishness and boorishness *versus* generosity and graciousness. The TV viewer is secretly entitled, but must be conciliatory, lest hosting become coercion or the individual be exposed as preferring TV to friends.

The underlying issue is, Who is accorded control and authority in the new zone of the TV set? The problem ramifies when the machine itself and the surrounding space come into play. Who has the prerogative to touch the machine? Who has the right to speak while a program airs? The host or hostess, owners of the property, of course, but in the 1950s, and again when color televisions appeared in the 1960s, regular tuning adjustment was likely to be necessary, and ownership and tuning skill were not synonymous. *The TV Jeebies* suggests several orders of violations and of resentment, first of the "amateur engineer" visitor with his blitz of technical terms and compulsive adjustment of the picture (his "expertise" presumably unearned because he owns no receiver himself). He transgresses in touching the object, manipulating its controls. So too the "station switcher" clicking from program to program, while the "screen hog," as the term implies, appropriates, possesses the picture (99–105). Other transgressors include the interruptive "conversationalist" and the "critic," both of whom violate the owners' sound space (and indicate how far the adult viewers were, at that time, from the dual, simultaneous attention to television and habitat eventually to become commonplace).

"I was allowed to make the image brighter and darker," writes the *New Yorker* TV diarist, "to make it slide off sidewise until it dissolved into a series of vertical lines" (44). He *was allowed*, he tells us pointedly, grateful to be given permission by the owner. And this is the point, that only the owner can designate or give permission for access to the machine and its space. Underlying the various offenses is the unarticulated premise that the rights of private property ought to govern control over the TV set and its surrounding spatial zone.

Naturalization

Soft Rounded Corners

The television receiver, named "monster" and The Westminster in the 1950s–1970s, emerges in the 1980s in very different terms:

> "We've been living together for a year now. She's shaken me awake and put me to sleep. The good news is she speaks many languages and is full of stories. The bad news: she often repeats herself. This baby makes a statement—she's a 27-inch TV." (Milward)

> Our TVs are active family members. They entertain us, teach our kids good stuff and bad, babysit in a pinch, give us the latest news, and, yes, drive us crazy. (Editorial, *Special Reports:* The Family 3)

The television set of the late 1980s appears in anthromorphic terms as live-in lover and family member. Even after decades of the many slang terms for television, most of them derogatory (e.g., idiot box, boob tube, electronic babysitter), these associations of eros and kinship are startling. Such self-consciously blatant connections between television and lovers or families is rare in the very extremity of the erotic and familial claims made. The statements flout anthropomorphism itself. (Both appeared in a publication of the Whittle Communications Corporation whose extensive telecommunications interests would promote endorsement of the idea of the television as intimate.) Still, the notion that the turned-on television set can be so personalized in contemporary American life suggests the degree to which television has become an integral part of the natural order. Having noticed the acculturation of the TV receiver as furniture, we turn now to the abstracted lineament of television, the screen, which after forty years has become the irreducible X of TV.

The round television screen was available in receivers until the early 1950s, when the slightly bulging rectangle with softly rounded corners won out to become the dominant screen configuration, and thereafter a generic logo for television itself. But a historical backward glance shows that the shape was by no means new in twentieth-century America. For one thing, the small-screen sets in huge cabinets were immediately reminiscent of the radio, as a story by Bobbie Ann Mason makes graphic. "The set . . . a ten-inch table model with an upholstered sound box in a rosewood cabinet . . . resembled our radio [and] for a long time I was confused, thinking that I would now be able to see all my favorite radio programs" (in *Shiloh*, 39). But the shape of the glowing glass of the radio dial was only a miniature of the rounded-corner rectangle widely deployed throughout the culture. Railroad car, trolley, and airplane windows had taken that shape, as did the photographic frame, both in still and motion picture photography. Before the TV set entered the domicile, family albums were pasted up with snapshots configured like the TV screen. And insofar as the television screen at first mimicked the frames of motion picture film and of train windows, the rectangle with soft rounded corners meant motion, travel, and entertainment from the beginning of the TV era. Advertisers exploited this connection to persuade sedentary viewers that they were really traveling activists—that as DuMont had proclaimed, the new "window on the world" would "pierce the barrier of distance." The success of this message may be inferred from a New York Central Railroad ad of 1956 showing two television actors, in real life a married couple, Peter Lynd Hayes and Mary Healey, facing out from the window of their 20th-Century Limited railroad car as it left the station, the railroad seeking to reclaim the vibrancy of the "TV stars" from the glass plate of the television screen to the plate glass of the railroad car (*The New Yorker*, 32 [27] [Aug. 25, 1956]: 71).

The coincident postwar fashion of the suburban picture window and the TV screen has been often remarked, with emphasis on the outward gaze through the vitreous medium into other environments beyond the household walls. But at the same time, the picture window furthered the association of the television screen as a part of the postwar American "utopian vision that included happiness as well as security" (May, 174–175).

As the two windows reinforced each other, certain furniture (e.g., the Stratolounger easy chair, the BarcaLounger, and the La-Z-Boy) was marketed explicitly for home viewing. The associations of the TV screen and the domestic sphere itself became reinforced—intimate yet insulated, familiar and safe. "No longer the talking box in the living room," writes a journalist, "TV has become instead a picture window overlooking a vast, interconnected video neighborhood that, for many of us, is Our Town" (Stone).

The journalist speaks explicitly about "our town" as the community network of viewers following soap opera or mini-series characters. But one implication of her statement concerns the encoding of meaning in the shape of the screen itself, regardless of what program might be playing or whether the receiver is turned off and the screen blank. Detractors of television can speak, as Reynolds Price does, of "the gorgon-eye of television," meaning petrification of the viewer and the physical repellence of the object itself (3). But visual texts of several kinds indicate that the soft rounded corners of the television screen have come to signify the safe insularity of the domestic realm, the "security" of

Microwave oven whose design mimics TV set, 1977.

the Cold War era. A historian of design observes that from the 1950s onward, "the shape of the tube burned itself into the subconscious, and soon it was reflected in automobile grilles, the windows of buildings, appliances, furniture, and decorative patterns" (Pulos, 299). These range, as we see, from costume jewelry to kitchen refrigerator magnets.

The TV screen commands attention in and of itself because, once advertisers and networks made it synonymous with safe insularity, other commercial interests could exploit it for that very quality. The TV screen, having acquired certain meanings, could in turn be deployed in other contexts precisely because it was assumed to bring those already secured meanings to them. Familiarized, it in turn familiarizes. Associated with the secure, the safe, it can be exploited precisely for those values. For instance, manufacturers of microwave ovens, working to overcome anxiety about low-level radiation in the nuclear age, have marketed these kitchen appliances in a design so analogous to the television set that the consumer's suspicion about radiation and its health hazards is abated, since the soft rounded corners of the lighted glass front suggest the safety and security of the habitat, signified in the TV screen.

The most revealing instance of the connotative power of the soft rounded corners lies in advertisements for air travel, especially in the 1960s, when the airlines, rapidly acquiring new fleets of jet aircraft, promoted recreational travel for couples and families. While the railroad advertiser sought to reglamorize a fading mode of transit by identifying it with TV star passengers, the airlines did virtually the opposite. Needing to overcome passenger anxiety, their advertising illustrators showed the airborne couple or family in an airliner interior whose seats looked like TV lounge chairs. Repeatedly in these ads, the airplane window is configured as a television screen to allay anxiety about flying. Of course, aeronautical engineers did not specify that airplane window designs mimic the TV screen. But the connotations of the screen itself were emphatically transferred to the airplane window in the age of jet passenger service. The safety and comfort of the living room or "family room" are emphatic in these ads, in which the soft rounded corners function as reassurance. To be thrust forward at six hundred miles per hour at a height of thirty thousand feet off the ground is a reality suppressed by these texts, in which the airplane interior is a living room with a TV screen. One writer, Michael J. Arlen, has made this connection. On a flight from Chicago to New York, he reported in 1978, "It occurred to me that there was a rather striking similarity between what I was experiencing . . . flying in a modern airliner, and what I've felt often happens as I watch television" ("The Air," 155). Arlen

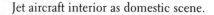

Jet aircraft interior as domestic scene.

lists the similarities as states of inescapable passivity, quietude, and inaction, and "non-aggression in contrast with the aggression of the 'outside world.'" The screen and lounge chair together encapsulate and insulate. The screen shape is culturally naturalized, no longer the face of the cathode ray tube but an inscription of safety and security to be deployed throughout the culture.

Television Lexicon

In 1948, H. L. Mencken wrote an essay on "video verbiage" in which this journalist and social critic, who had undertaken a two-volume study of the development of American English, mused about the new terms entering the language as a result of television. Just as movies had borrowed a technical vocabulary from the stage, and radio in turn "filched words right and left from the movies," so television, in the late 1940s, he noticed, was borrowing heavily from its own antecedents. Mencken observed that the film terms, *angle shot*, *frame*, and *white meat* (slang for an actress) were current in television usage, but he wondered which other terms, generated by the new medium, would survive and be naturalized into the American language.

Video for the picture and *audio* for the sound seemed promising in 1948, although the prefatory *tele* also seemed a term likely to endure. But "what are the fans to be called?" asked Mencken. *Televiewers?* or simply the truncated *viewers?* Or would *gazer, looker*, or *looker-in*, these last two terms in use from the radio antecedent of "listen in," become the preferred and commonplace designations (Mencken, 112–14)? Personally, the ascerbic Mencken liked *gawk*, but he went on to speculate that other words such as *telegenic* and *telecast* would prove durable in the new television age. In the retrospect of some forty years, some of these terms seem silly, as if the long-discarded phrases (*looker-in!*) would naturally be discarded, the linguistic survivors virtually destined to take their place in the lexicon of a television-era culture (the very term, television, part of this linguistic destiny). In this sense, the topic of TV language even seems quaint ("horseless carriage" may come to mind, and "roadster").

The naturalization of the TV lexicon may seem a foregone conclusion, but less obvious is the way in which terminology from television structures experiences in the non-TV realm. Terms like "TV" or "telecast" long ago disappeared from self-consciousness and vanished into the common fabric of everyday language, but television has continued nonetheless to provide the terms that give shape to experience ostensibly unrelated to television.

No sooner had Mencken described the exotic neologisms of the new age of television than disparate texts began to show the ways in which non-TV life was beginning to imitate television. The captions in cartoon humor indicate a growing governance of television over diverse areas of human experience and situations. The formats and lexicon of the realm of television have become the behavioral model for many moments in personal and domestic life, the humor cartoons suggest.

In drawings, for instance, featuring a young woman and man entering the living room to announce their engagement ("We interrupt this program for an important news bulletin"), or showing a father just home from work ("Yes, it's Daddy—live and in color"), or an automobile outing in which father and children cope with a flat tire in the rain ("Don't you understand? This is *life.* . . . We can't switch to another channel"), or an unruly mealtime scene at the dining room table—the humor in all of these and others turns on the appropriation of TV terms for personal situations. The captions come directly from television: "Wake up, Fred! This is prime evening time!"; "Dad, what category should I go into when I grow up?"; "Ready for an in-depth, on-the-spot, first person overview of my day?"; "I'll make a deal with you I'll try to be more like Harriet if you'll try to be more like Ozzie." The movies, *Splash* (1984) and *Earth Girls Are Easy* (1989), like the cartoons, show television at work in the structuring of language when the aliens learn English from television. The *Splash* mermaid, played by Daryl Hannah, watches television on Bloomingdale's sales floor for six hours, then says to her boyfriend (Tom Hanks), "Now I can ask you questions, and if you answer them correctly, you can win one of these valuable prizes." And television's structuring of thought need not be verbal. In the early 1990s, a cartoon drawing shows a middle-class family of three at dinner, their thought-balloons showing they secretly think themselves a version of The Simpsons. These drawings and captions presume familiarity with the conventions of television and its programs, from news to sitcoms. The humor pivots on the initial incongruity of each application of television to the personal context—then sustains itself on the self-conscious surprise at the validity of the congruence. The TV term is natural within the on-screen TV environment. When applied to situations in the off-screen world, the TV term appropriates those situations and refocuses them as parodies. What is parodied, however, is not television but non-TV personal and social life. (*The New Yorker*, 32 [March 24, 1956]: 37; [December 8, 1956]: 48; 36 [March 19, 1960]: 119; 41 [July 24, 1965]: 58; 42 [May 2, 1970]: 42; 46 [July 25, 1970]: 21; 52 [Oct. 4, 1976]: 45; 66 [June 11, 1990]: 33).

Educational settings also show TV structures of language, from elementary schools in which learning games mimic programs like "Wheel of Fortune" or "Jeopardy," to the highest levels of the academy. In 1990, the feminist scholars Sandra Gilbert and Susan Gubar presented a lecture entitled "Masterpiece Theatre: An Academic Melodrama," in which the audience were invited to "pick up the remote" and "flick on the TV" for their "docudrama" presentation on the vexed situation of literary theory in the humanities. The lecture is structured on television's "split screen" and "the TV screen becoming [a] computer screen" and at one point "cut[s] to a television studio" (1, 5, 8, 12, 39). The academic, professorial discourse is itself structured on the formats and language of television.

Recent fiction also exploits the lexicon and the forms of television, in so doing legitimating the televisual as the primary—in essence, the natural— structure of experience, including that of the language in which it is cast. Television becomes the writer's phrase book, which is interesting because, unlike the cartoon caption whose TV term assigns total meaning to the illustration, TV phrases enter the story world desultorily, not as part of a deliberate overall design but as the common lingua franca. A father checks on his sleeping children: "I looked for a blanket to adjust, a toy to remove from a child's warm grasp, feeling I'd wandered into a TV moment" (DeLillo, *White Noise*, 244). A woman looks outside and "watches storm warnings roll like credits along the bottom of the television screen" (Agee, 2). A mystery novel shows onetime high schoolers reminiscing about "which one of the men from U.N.C.L.E. was the cutest, David McCallum or Robert Vaughn," and in another novel, a child's fantasy of a reunited family is cast in terms of a nighttime TV show: "The whole family was up there, all spread out like the little winking lights of the "Tonight Show" skyline" (McCrumb, 238; Wolitzer, 13).

Characters, too, can be described in the language of their on-screen TV images: "She had the opaque look of a hair-curled woman on the evening news" (*White Noise*, 152). A man's boss looks like "a TV talk-show host" (Turow, 10). A neighborhood party becomes "like one of those game shows" (Carver, *Cathedral*, 197). And this kind of referencing holds for other recent fictional texts, featuring such shows as Ed Sullivan's "The Toast of the Town," "I Love Lucy," "Jeopardy," "The David Letterman Show," Christmas Specials with Andy Williams, Perry Como and Marie Osmond. All these and other programs become integral to the structures of story-telling as contemporary narrative shows its dependence on the structures of television. (See Donald Barthelme, "And Now Let's Hear It for The Ed Sullivan Show" [*Guilty Pleasures*], Trey Ellis, *Platitudes*, Peter Cameron, "Odd Jobs," [*One Way or Another*], David Foster Wallace, "Little Expressionless Animals" and "My Appearance" [*Girl with Curious Hair*].)

Even personal crisis is shown in relation to television. A boy caught in his parents' breakup retreats in fantasy to his private TV show, "The Perfect Brothers Show" with its theme song, emcee, and weekly guest star—a fantasy in which television becomes the new, modern fairy tale (Leavitt, "Danny in Transit," *Family Dancing*, 107). "We only *think* we're alive," cries a desper-

ate furnace repairman in Banks's *Continental Drift*. "We watch that fucking
TV screen, and we think we're like those people there, fucking Hart and Hart
[the glamorous husband–wife detective team], and that makes us forget that
we're not like those people at all. We're dead. They're pretty pictures. We're
dead people" (Banks, *Continental Drift*, 31).

Television shapes language, fiction tells us, even at the most painful
moments of human experience, forcing the writer to contend with it and even
capitulate to it. In Raymond Carver's "A Small, Good Thing," a young
mother leaves the hospital in which her son has just died unexpectedly:

> "No, no" she said. "I can't leave him here, no." She heard herself say that and
> thought how unfair it was that the only words that came out were the sort of
> words used on TV shows where the people were stunned by violent or sudden
> deaths. She wanted her words to be her own.
> "No," she said. (*Cathedral*, 81)

But the shocked, grieving young mother does not find a personal language
different from television. This moment, like the others in television-era
fiction, is perceptually governed by television, its optical shadings, its programs, its stock scenes, and character types and rituals as television serves the
writers' stock-in-trade of exposition, plotting, frame of reference, point of
view, and so forth. At times critical of television, these writers nonetheless
exploit it, knowing it to be the contemporary environment, the cultural matrix
shared by writer, reader, and characters.

Television, then, becomes the standard by which the non-TV world is
perceived, ordered, and understood. It becomes a kind of parent language,
furnishing the governing terms by which non-TV phenomena can be expressed and grasped. The frame of reference for virtually every life experience
is television, even if an author is guarded against television's cliches, as is
Carver—or critical of TV, as Turow's narrator is, complaining that "TV and
the movies have spoiled the most intimate moments of our lives" by replacing
spontaneity with "conventions which dominate expectations" (84). (Complaint aside, Turow uses the shortcut of a TV simile for the thumbnail
portrait—the boss as "a TV talk-show host"—an image that every reader will
grasp immediately.)

Defamiliar

A discussion such as this one, conducted in a kind of structured diptych,
hinges television's cultural initiation against its naturalization, yet requires
one more panel. This last is a recognition of the most recent, and related,
exploitation of television. I speak of its deliberate defamiliarization. To move

through the two categories, the initiation and the naturalization, is not quite sufficient. There is another, subsequent and perhaps inevitable part of the acculturation of television. Here, we recognize the act deliberately intended to make television conspicuous once again. Having vanished from everyday consciousness in its very omnipresence, television, meaning the receiver and the viewer's relation to it, stands ready to be presented once again to consciousness. Since the familiar cannot be made unfamiliar (disguise, based on concealment, being something else entirely), the TV text that works to make television newly visible must defamiliarize it. Television itself does some of this. In "Moonlighting," Bruce Willis, playing David Addison, has occasionally broken the theatrically conventional fourth wall and addressed viewers directly, explicitly reminding them that the program is television. And CNN's newscasts are allowing into the audio portion certain technical cues customarily not broadcast to viewers. In both instances, the viewer suddenly becomes an insider, and the production is revealed to be a process and a construct as well as a product. Defamiliarized television, then, commands attention anew. The point is to refresh or revivify one's experience of it, to raise the mundane to a new level of awareness.

As a point of departure here, let us consider texts that might at first seem to defamiliarize television but in fact do not do so. They are both examples of parody. By 1955, a cartoonist presented James McNeil Whistler's *Arrangement in Gray Black No. 1: The Artist's Mother*, better known as "Whistler's Mother" as the elderly woman sits in her nineteenth-century dark day dress with cap—but this time posed before a TV set. The painting, like Leonardo's *Gioconda* (the "Mona Lisa") has been so often reproduced that one is virtually unable to see it freshly. The cartoonist relies on this overfamiliarity and, adding the television set, reconfigures the painting, in fact defamiliarizes the time-worn portrait. In the cartoon, television is the device enabling the painting to be defamiliarized.

Similarly, in 1963, a verse parody of *The Waste Land* appeared, based entirely on television, and taking as its point of departure the much-publicized criticism voiced by the then-Federal Communications Commissioner, Newton Minnow, who in 1961 called television a vast wasteland. "Suppose that the poet [T. S.] Eliot were confronted by an American television set," said the parodist. "What might he have to say then?" (Armstrong, 92). The reader proceeds immmediately to the parody:

> April is the cruelest month, breeding
> Cowboys out of the dead land, mixing
> pilot films, stirring
> Tonight Show with Today.

Winter kept us warm, covering
Sullivan with Walt Disney, feeding
Lawrence Welk with Mister Ed.
Summer surprised us, coming over the coaxial
With a shower of Talent Scouts; we stopped on Mitch Miller,
And went on to midnight, into the Paargarten,
And drank coffee, and watched for an hour.

The verse continues through parodic versions of the best-known passages of Eliot's poem, the basis of complaint about this April TV wasteland being that the seasonal reruns have begun. The parody is filled with names of programs and television figures from *The Price Is Right* to Yogi Bear. Of course, the premise of this parody is the reader's familiarity with T. S. Eliot's poem, taught so widely in the college classroom that anyone who had studied modernist literature would be expected to know it, to have some lines memorized. Once again, it is this familiarity on which the parodist relies. And in fact it is *The Waste Land* that is defamiliarized in the parody. Once again, television and its lore are the instrument by which the modernist classic is able to be seen anew. Should one object to the trivialization of a classic, the point remains that it is the poem that is reconfigured, not television.

But defamiliarized television works in reverse. In this latter case, the artist uses some instrument as a means by which to defamiliarize television. One example of a visual text shows this. It is a drawing in David Macaulay's *Motel of the Mysteries* (1979), an illustrated narrative based on the presumption that in the 1980s, the North American continent virtually drowns in its own detritus and that its civilization is excavated by archeologists two millennia later. At that time, the scientists unearth a motel, which they construe as a sacred site, and Macaulay plays fast and free with a range of "artifacts," from the do-not-disturb sign ("the sacred seal") to the toilet brush ("the sacred aspergillum"). As one might imagine, any number of late twentieth-century commonplace objects are defamiliarized in *Motel of the Mysteries*. As for television, it is discovered in a room with the skeletal remains of a male *Homo sapiens* lying supine on a bed before it, a remote channel changer clutched in hand, or flanges. The television set is, the archaeologists decide, The Great Altar, a "magnificent structure toward which everything in the outer chamber was directed" and "represents the essence of religious communication as practiced by the ancient North Americans. . . . Communication with the altar was symbolically continued into eternal life by placing a communicator box in the hand of the deceased" (28–29, 56).

David Macaulay uses the ploy of the archaeological expedition into a lost civilization as the instrument by which he defamiliarizes television. The satire

TV defamiliarized as "The Great Altar" in David Macaulay's *Motel of the Mysteries*. Courtesy David Macaulay and Houghton Mifflin, Inc. Reprinted by permission.

on "The Great Altar" is a barb at the excessive importance he feels is accorded to television. In his drawings and in his narrative the ordinary television receiver is intact, yet seen anew because it is defamiliarized.

The drawings and text by Macaulay measure how deeply television has embedded itself in American culture. To raise it to a new visibility, he must move his readers nearly two thousand years into the future. The text, in fact, repositions television in a new chronological time line. If it has taken some forty years of cabinetry design, together with social adaptation in custom and language to make television as invisible as it is omnipresent, *Motel of the Mysteries* tells us that an imaginary two millennia are necessary to bring TV into social focus once again.

And a new, defamiliarized focus on television is precisely the objective of the chapters that follow.

Electronic Hearth

The television set dominated the living room in most homes.
. . . In many cases the warm flicker of the fireplace was preempted
by the television's electronic glow.
> Arthur J. Pulos, *The American Design Adventure*, 1988

So they sat . . . very much at home. Very much at ease. . . .
They watched the soundless images on TV as if they were logs on a
hearth.
> Stanley Elkin, *The Magic Kingdom*, 1985

Our yule log video brings you the festive warmth of a glowing
fireplace on any TV set or VCR . . . 60 minutes of a warm,
crackling fire.
> *Bloomingdale's By Mail*, Christmas Catalog, 1989

Network television featured the new electronic hearth all winter long, Monday through Friday of 1989–1991 when the studio set of ABC's *Good Morning America* was constructed to resemble a living room with a crackling fire and a cracking big-screen monitor. Viewers saw both the open-fire hearth and the adjacent big color TV on which the host and hostess, Joan London and Charles Gibson (as some have noticed, surrogate parents), conducted interviews by electronic hookup.

The television was not pushed up against the fireplace to save room in a crowded studio. The hearth and the color monitor were strategically positioned side-by-side to convey a distinct message. Viewers were asked to grasp and to accept the analogy of the hearth with the television, and in turn to associate their own TV sets with a glowing hearth.

Any toddler could distinguish the television from the fireplace, of course, but ABC's conflation of the two objects intentionally makes it impossible to separate the implied meanings of the one from the other. On the contrary, it serves the interests of the network and its advertisers to blur the lines between them and make distinctions of meaning impossible. By design, *Good Morning America* viewers are to see the two objects—television and hearth—as inter-changeable. The two are presented as analogues. More than that, they are past and present forms of each other, and all the meanings contained in the hearth are intended to transfer to the TV receiver. All the historically accumulated associations of the American domestic hearth are reproduced in the television.

The *Good Morning America* studio set reminds us that technologies can be bearers of ideological values carried forward from the past into the present. The reverse is often argued, that new technologies enter the culture in interven-tionist, disruptive ways, and are, therefore, socioculturally discontinuous. Yet, these same technologies are presented to consumers literally in terms of the past, of what is familiar in function and meaning. (The horseless carriage preceded the automobile, and still, after a century, we measure the engine's energy output in equine terms.) Technology can be disseminated with its value system intact, bringing ideology forward, assuring the public of cultural continuity, conveying the message that there is no fundamental change, that history itself is a fiction.

The TV hearth makes a good case study, showing the extent to which the TV environment is laden with values and traditions that carry forward from a preceding period. Equally important, to notice the meanings embedded in the hearth from colonial America to the television age is to see that certain interests are served by pressing an ideological program centered on patriotism and domestic security. The hearth may seem politically neutral, a primitive heating system long superseded by closed stoves and furnaces and, in the age of central heat, sustained in architectural decorative arts. But, in fact, the American open fireplace is freighted with carefully inculcated meanings adhering through the processes of history and strategically deployed by corporate and media interests in the TV receiver. We err if we fail to disclose them and to consider why it is that TV viewers have been encouraged to experience the cathode tube as an electronic hearth.

The connection was already in place as early as 1944, when an Alan B. DuMont Corporation booklet stated that "scientific research brings television to brighten your fireside" ("Times, Tubes and Television"). The corporate advertiser carefully emphasized the enhancement, the "brightening," of the fireside, not its replacement by a new technology, though four years later, in 1948, a journalist with a more disinterested view saw the two entities as antitheses and competitors and asked, "What will television do to that warm old native-stone tradition of the American home, the fireplace?" The journal-ist, positioning hearth and television in mutual exclusion, worried that "there

Saturday Evening Post cover, 1954. Courtesy © Curtis Publishing Co.

Drawing from *Saturday Evening Post*. Courtesy © Curtis Publishing Co.

won't be room in most small living rooms for both fireplaces and television sets, and the fireplace probably will have to give way." A year later, a business journalist said it had already given way, that "people are recessing their sets in the fireplace to save room" and that "the fireplace . . . is on the way out." In 1950, another lamented, "Fireplaces are losing out to give their preferred position to the TV set." These statements are tinged with ambivalence over loss and displacement of the hearth by the television ("Hypnosis in Your Living Room," 71; Manchester, 43; "TV: An Interim Summing-Up," 7).

Subsequent writers, however, have overcome ambivalence by changing binary opposition into a merging of the two objects, which is the very strategy used by the ABC set designers. Assenting to the consumer culture's program of planned obsolescence, they proclaim television to be the newer version of the older hearth, something like a more recent model. This hearth–television relation allows the continuing survival, intact, of the hearth within the TV receiver; instead of either/or, we have both/and—and so in 1989, a newspaper journalist writes, "In the home, the television has become an electronic hearth. It has brought the American family back together" (York). The historian of television, Erik Barnouw, elaborates in his history of the medium: "For several decades, patterns made in a tube by fusillades of electrons have played an extraordinary role in our lives. The glistening mosaic, replacing the hearth, has become the focal point of the home. In living color more real than life, the swirling dots represent the world; they tell us of the good, great, beautiful, and desirable" ("Foreward"). In 1989, in *Life* magazine's 50th anniversary celebration of television, the media critic, William Henry III, similarly observes that "television is now the hearth around which we meet, the cool fire of domestic evenings and the tribal gathering place for moments of ritual celebration or mourning" (66). In this sense, the television is a replacement of the hearth, a new positioning. The hearth does not disappear, as the 1940s critic had feared, but is rematerialized and renewed in the TV set.

From the late 1940s, the TV hearth has also been represented as accommodating the community, since television was often portrayed as a quasi-theatrical experience in the home, with friends and neighbors invited in to view certain programs, perhaps championship boxing or the *Texaco Star Theatre* with Milton Berle. Cartoons showed the rearranged living room filled with household chairs or stacking viewers on a stairway in simulation of the raked theatre floor. These drawings derive humor precisely from the conventions of the theatre or of arena sports now enacted in the home with friends and neighbors. Above all, this postwar era was declared in the catch phrase, "togetherness." The nuclear family, male and female, were to work for, aspire to, the American bonding and unity expressed as "togetherness," an idealized state fully compatible with the TV hearth (*The New Yorker*, 25 [25] [Aug. 13, 1949]: 18; [30] [Sept. 17, 1949]: 33).

The novelist John Updike has drawn the hearth–television analogy explicitly in the novel, *Roger's Version* (1986), in which a boy, Richie, does his homework: "Why, with a living room, a library, and his own good-size bedroom at his disposal, Richie insisted on doing his homework on the very surface where his mother was trying to place mats and dinner plates, while a ten-inch Sony crackled and chattered not a foot from his face, I couldn't imagine. Or so I said . . . Of course, I secretly understood: the primitive appeal of the hearth. Television is—its irresistible charm—a fire. Entering an empty room, we turn it on, and a talking face flares into being" (34–35).

As the electronic hearth, television is emphatically joined to American history. The discourses of corporate advertisers, media interests, and consensus journalists all evoke in the very term—hearth—the traditionalism of the past. Therefore, television can be claimed as the newest embodiment of values that go deep into the national culture as that culture is historically represented, say, to school students and to an adult public assumed to be middle class in outlook and material means. To analyze the TV environment is necessarily to pursue these antecedent meanings, these layers of value and feeling claimed for the TV age.

Since the mid-nineteenth century, the hearth has figured persistently in the major texts of canonical writers. The open hearth–fireplace elicits a cluster of meanings beginning with domesticity. In *Little Women*, Louisa May Alcott's March girls survey the neighbor's house to find "best of all, a great open fireplace with quaint tiles all around it" (65). One anonymous magazine writer recalled "the wide, open fireplace" in which "great back-logs dozed . . . and smaller sticks broke and crackled before them, and high over the flames a line of pots and kettles swung from the crane." It was in those days "a delight . . . to sit near the chimney . . . and see the shine of the great fire playing over the burnished tops of the andirons" ("Grandmother's Houses," 116).

Harriet Beecher Stowe was one of several writers who represented the hearth as central to American history and the national ideals of plenitude, domesticity, and courage. In *Uncle Tom's Cabin*, she presents the kitchen of Aunt Chloe, the slave cook of the Shelby family, as an open-hearth cornucopia. "It is her you see by the fire, presiding" Stowe says to the reader, emphasizing that Aunt Chloe is the embodiment of "satisfaction and contentment . . . a cook in the very bone and centre of her soul" (24). As Chloe tends her baking pan, Stowe leaves her to "finish the picture" of Uncle Tom's "cottage," and readers next are shown a bedroom corner turned "*drawing-room*," also with hearth, this a fireplace on whose wall hang "scriptural prints, and a portrait of General Washington" (24–25). Stowe gives readers two hearths, two shrines, one to domesticity, the other to patriotism, the very founding of the nation. The two are complements, each necessary to complete the other.

The domestic values of the open hearth extend to the patriarchal origins of the nation, as Stowe asks what motivated "our Revolutionary fathers" to go "barefooted and bleeding over snows":

> It was the memory of the great open kitchen fire, with its bag-log and fore-stick of cord wood, its roaring, hilarious voice of invitation, its dancing tongues of flame, that called them through the snows of that dreadful winter to keep up their courage, that made their hearts warm and bright with a thousand reflected memories. (*House and Home Papers*, 7)

Here the open fire is both the signifier of wholesome festivity and, more important, the motive for patriotic endurance. Without the hearth, the

American Revolution would not have been fought and could not have been won. And in the era of the Colonial Revival, beginning at about 1865 and extending into the 1920s, the open fire continued to be a central icon. "The open fire shall glow again," said one commentator. "No one who has ever stood before the old-time, spacious fireplace . . . can have failed to feel a thrill at the experience. Why, here Washington lived. And this is the kind of fire he had! . . . What an atmosphere of old-time ways and old-time living—the very heart of the republic, even now" (Huston, 53).

The historian of American popular culture, Karal Ann Marling, notes that illustrations of Betsy Ross sewing the flag usually featured a "capacious hearth," and that the "ideal colonial hearth needed a man of the house," customarily George Washington, "and a lady with some active claim to the respect of history" (18–19). For that matter, the hearth could become the shrine of Revolutionary heroics, as in the Connecticut State Fair of 1876 in which a portrait of the state's Revolutionary War hero, General Israel Putnam, was mounted above a "fire-place with great brass andirons" (Marling, 35). In the 1920s, Wallace Nutting, the popular exponent of early American furniture, extolled the "hearth" as a term reaching "to the roots of our language," denoting "warmth and protection," "sense of home," and intergenerational bonding in which boys learned "the lore of the tribe," while their grandfathers sat "in the chimney corner, fondling their grandchildren upon their knees" (Marling, 175). From architecture to silver design to projects like Colonial Williamsburg or Henry Ford's renovation of the Revolutionary-era Wayside Inn in Sudbury, Massachusetts, the enduring popularity of the Colonial Revival meant the concurrent reverence for the hearth, emphatic in its prominence in state and national expositions, and in the zest with which middle-class Americans installed open-hearth fireplaces in houses heated by furnaces and closed stoves.

In all these passages, the recurrent themes of domesticity with family centeredness couple with patriotism (a particularly male virtue in its association with Washington and the Revolutionary Fathers) to make the hearth serve two gender-roles. The hearth embraces male and female realms, serving the women's sphere of family-centered domesticity as well as the masculine forms of patriotic militarism and nationalism. (Even the 1960s–1970s would echo the masculine form of nationalism as the post-presidential election of Richard Nixon brought the commander-in-chief and his lieutenants offices "with antique desks . . . and fireplaces that were kept burning year-round" [Hersh, 11].) But both sexes can claim the hearth as their own. And one would be neglectful not to cite the best-known literary text on the hearth, John Greenleaf Whittier's poem, *Snow-Bound*, in which the New England family gather at the central chimney's open hearth when a winter storm isolates them from the outside world. The fire, Whittier writes,

> Burst, flower-like, into rosy bloom
> ....
> Our own warm hearth seemed blazing free.

The poem goes on to describe the cat and dog lying in temporary peace at hearthside (invoking the millennial peaceable kingdom), the mug of cider between the andirons, the bowl of nuts, the "hearth-fire's ruddy glow." And later we learn what it means, this hearth: "food and shelter, warmth and health, / And love's contentment more than wealth." This popular poem, written in 1866 as a retrospective about 1820s America, and revised just before Whittier's death in 1892, was a landmark text of the Colonial Revival. It was published in front-parlor editions and helped to give a school of American poets, including Henry Wadsworth Longfellow and Oliver Wendell Holmes, their name as The Fireside Poets.

And Longfellow, like Whittier, exploited the colonial-era open hearth, "the blazing fire of wood" helpful particularly to him in framing his *Tales of a Wayside Inn*, which was based upon Boccaccio's *Decameron* and Chaucer's *Canterbury Tales*, and featured a group of travelers who gather around the parlor fire at the Sudbury, Massachusetts, hostelry in preindustrial America:

> One Autumn night, in Sudbury town,
> Across the meadows bare and brown,
> The windows of the wayside inn
> Gleamed red with fire-light through the leaves
>
> ....
> The fire-light, shedding over all
> The splendor of its ruddy glow,
> Filled the whole parlor large and low. (232–33)

Longfellow thus launches his tales, mixing Nordic sagas with quasi-medieval legends, yet including American history as the Wayside Inn landlord tells of "Paul Revere's Ride," since memorized by generations of U.S. school children and, in the *Tales*, bonding the glowing open hearth of the Wayside Inn to the patriotic values of the American Revolution.

Though the Fireside poets were subsequently dismissed as merely genteel by the Cold War-era critics, the valorized American Romantics, too, incorporated the hearth into their own ethos of romantic literary theory. Nathaniel Hawthorne wrote of the "vivifying element . . . firelight" in a sketch entitled "Fire Worship" (501) and, in *The Scarlet Letter*, wrote of the "dim coal-

fire . . . the half-extinguished anthracite . . . [that] throws its unobtru-
sive tinge around the room . . . and we behold . . . [at] one remove
farther from the actual, and nearer to the imaginative" (66). In romantic
terms, the fireplace is psychologically transformative. The hearth is accorded
privileged status as a vehicle of transcendence doubtless more acceptable to
middle-class readers than were the transformative media of distilled spirits and
the opium pipe.

All this, then, can be tapped for television discourses promoting ideas on
domesticity and patriotism in the postwar United States. And the TV hearth
comes by way of the radio hearth, too, with its association of friendliness and
familial cohesiveness associated with all of broadcasting in the United States.
One recent writer glimpses postwar suburban life in which one could "watch
increasingly larger television sets in the 'TV room,' where all the furniture was
bunched cozily around the Magnavox" (Rudnick, 181). Popular visual images
of the early 1950s capture the nuclear family gathered to watch *Kukla, Fran
and Ollie* in their living room, forming a semicircle of mom and dad, son and
daughter, the boy "roughing it" on the rug, though the whole family is well
groomed and respectable, and prosperous in middle-class terms. Atop their
TV cabinet, we see, they have put a clock, perhaps to remind them of the hour
when the program changes, but more likely of the children's bedtimes and the
adults' too, since father and mother must not allow television to distract them
from scheduled commitments. The clock tells us the family will not be led
astray by their new TV set, just as clothes tell us they could, if they chose, go
out of the home at any moment, to pay an evening call or to dine. Television
will not tempt them into slovenly domestic ways. Yet they have, appropriately,
chosen to gather at this new hearth. Their profiles tell us of their attention and
satisfaction, and in fact this TV hearth would lose its meaning were the family
to be absent from the picture. The human figures say that the new cathode
tube hearth brings the family together in a scene of harmony and affection.

These meanings are presumed by the 1960s, when RCA advertised its color
sets in a wood-beamed living room without human figures. Though no one sits
on the floral sofas, we are not to imagine their departure from the TV hearth—
but instead our invitation to gather there, to claim our place. America's "finest
homes," the ad says, are those in which the TV set is identified explicitly as a
hearth, its very cabinetry an "Early American" model named The Quincy,
reference both to a presidential family and their colonial New England origins
in Massachusetts. In the ad, the grandfather clock and family portraits of
ancestors in the background indicate tradition and culture, and family
continuity in this age of technological advance. This hearth-room is at once
feminine in the sofa fabric (chintz?) and masculine in the exposed beams of the
ceiling and rough stone chimney of a lodge. The ad copy informs the reader

Taproom at Wayside Inn, Sudbury, Massachusetts, memorialized by Henry Wadsworth Longfellow in *Tales of a Wayside Inn*. Courtesy the Wayside Inn, Sudbury, Massachusetts. Reprinted by permission.

Representation of white, middle-class American family of the 1950s gathered around the electronic hearth. Courtesy Quality Post Cards, San Francisco. Reprinted by permission.

In America's finest homes...
true-to-life color with RCA Victor solid copper circuit dependabil

RCA advertisement of 1960s.

that this interior is an actual home of a prominent family, and therefore certified as authentic. (Readers of the caption learn that it is actually the home of the actress Eve Arden, who starred in the television sitcom *Our Miss Brooks*, which made a successful transition from radio to TV.)

Central to the image is the dark fireplace that seems linked to the dark screen. Yet on closer scrutiny, the television is really seen to be off the boundary of the living room rug, and for the onlooker this creates a certain psychological tension. Visually, we wish to unify the room, and the ad layout guides us with the angle formed by the chimney/hearth rug/TV cabinet edges. Visually, the onlooker slides on the rails of that central angle, first to pull the entire scene forward toward the TV set until the chimney is so close that it frames the television screen and integrates it into the hearth and chimney. Or we can visually push the television deeper into the room, again to the place where it belongs, at the very hearth and right into the fireplace. Either way, this advertising image tells us the television set is to be the *real* fireplace.

And if we should imagine turning the television around, facing it into the room, the television and the hearth will mirror each other. The painted faces on the ancestral portraits, images themselves, in effect will "watch" the television images and in so doing legitimate the new electronic medium. We, ourselves, in any case, will take our own places on the parallel sofas, but clearly to watch the televisual fire of warm and vibrant color, not that nearly blackened stone cavity in which the embers are dying at this very moment. The color is already on-screen, inscribed in the word, C-O-L-O-R, in vibrant chromatics. We read the word, and read as well the on-screen colors that accord so well with those of the furniture in the room. The new RCA color TV will really give us, the ad says, the whole room and our rightful hearth, the colorful screen.

Presented to the public as a hearth, then, television assures the consumer that this technology, including color, does not disrupt tradition but acts, on the contrary, as an agent of continuity in an America whose values remain intact over centuries. RCA's "solid copper circuitry" is as "solid" as the stone and the heavy hewn beams, all durable materials from the North American continent supporting a durable image of the national values.

Yet the hearth carries other ideological baggage not acknowledged in these century-long images and certainly suppressed in the television-era discourses on patriotism and domesticity. True, the icon of the hearth embodies values extending to the founding of the nation and, since then, represented to be timeless and unassailable. Assigned meanings on patriotism, abundance, family cohesiveness, domesticity—these, centered in the hearth, transcend historical flux and take precedence over it. But when the hearth itself is examined as an object with a history, that is, as an object-turned-icon in the

Television in "Early American" cabinetry.

process of recurrent interpretation by poets, polemicists, exposition entrepreneurs, and others, then additional issues emerge. The very history of the development of the hearth from heat source to icon of the Colonial Revival extending to the television age reveals a cluster of deep cultural anxieties in the United States. The history of the hearth, including the televisual hearth, shows that the ways in which a national icon can come into being precisely in response to historical change, and can be strengthened exactly at those points at which social vulnerability seems greatest.

It is well to remember that the Colonial Revival itself occurred at a period of rapid, disruptive social change in the United States. Whittier wrote *Snow-Bound* in the aftermath of the upheaval of the Civil War, and several historians of the movement have pointed out the ways in which it flourished at cross purposes with urbanization, industrialization, and immigration, the salient sociocultural developments in the nation from the 1870s into the twentieth century. As the art historian Vincent Scully remarks, the Colonial Revival meant an escape from "an industrial civilization grown complex and brutal, from cities grown too dense and hard" (Rhoads, *The Colonial Revival*, 395). Frank Lloyd Wright recalled his entry as a young man into a Chicago of "torrential noise," "clamor!," "streams of human beings in seeming confusion." It was "terrible! This grinding and piling up of blind forces. If there was logic here who could grasp it?" (*An Autobiography*, 87–88).

As metropolitan life established itself with the growth of cities, the colonial way of life allayed anxieties about "an alienating and impersonal urban present" (Betsky, 243). When American men engaged in industrial work, the hearth (often with a gun over the mantle) signified individual masculine bravery harkening back to Colonial days (248). As hordes of immigrants descended with their own languages, politics, and traditions (the foreign-born population of the United States more than doubling between 1880 and 1930, from 6.7 to 14.2 million), xenophobia rose among "Americans whose ancestors had arrived earlier [and] were often fearful of the destruction of their own values and customs" (Rhoads, *The Colonial Revival*, 341).

This was the era of the formation of the Daughters of the American Revolution, of colonial-style architecture—at least of colonial facades—on buildings constructed for the schooling of immigrants, and of such rituals as Henry Ford's "Ford English School Melting Pot" pageant, "in which immigrants in native costumes descended into a great pot, while 'new Americans' emerged wearing business suits and carrying American flags" (Rhoads, *The Colonial Revival*, 357).

Another glance at Longfellow's *Tales of the Wayside Inn* proves useful here, for Henry Ford initially decided to construct a colonial American village on the site of the 1702 tavern–inn that was frequented by Revolutionary War soldiers and memorialized in Longfellow's verse. To construct the environment of the colonial American past on the grounds of the inn, located on the Boston Post Road, the automobile manufacturer shipped and reassembled an elementary school from Sterling, Massachusetts, and had built a working reproduction of an eighteenth-century grist mill, together with a reproduction of a New England church, which he named in honor of his mother and mother-in-law. While the Ford automobile precipitated revolutionary change in industrialization, transportation, and social customs in the United States, Ford himself exerted a major effort to enshrine an idealized Colonial past in a constructed material environment.

Throughout the twentieth century, when rapid social change prompted one hymn to "a return to the simpler ways of the forefathers of our country," one could cite the familiar political benchmarks of turbulence—Sacco-Vanzetti, the Bolshevik scare, the Palmer Mitchell raids, the Smith Act, violent and widely publicized labor strikes—all fostering the unabated continuation of the Colonial Revival (Rhoads, *The Colonial Revival*, 399). Within the American home, too, the hearth was an iconic bulwark against the dispersal of family members when central heating and electrification permitted individuals to seek privacy in separate rooms, and the telephone and automobile enabled psychological and actual transport away from the former family center as those once snow-bound now knew no bounds. The twentieth-century anxiety about losing one's grown children to the city or other migratory destination surely played a part in the promulgation of the colonial hearth.

Surveying the movement from the vantage point of American decorative arts, one historian remarks that "the tradition of the old has led to a continuing conservatism in the . . . arts in America that is quite at odds with the technological achievements of the twentieth century" (Carpenter, 158). As architect, for instance, Frank Lloyd Wright exploited the technology of new building materials, but from his turn-of-the-century Prairie houses outside the clamorous city of Chicago, up to Falling Water, the house he designed for the Kaufman family in the 1930s, Wright counted some three thousand fireplaces in his designs. "The *integral* fireplace became an important part of the building itself in the houses . . . out there on the prairie. . . . It comforted me to see a fire burning deep in the solid masonry of the house itself" (*An Autobiography*, 165, 523). During the 1930s, too, President Franklin D. Roosevelt intended to comfort and reassure Depression-struck America with his "Fireside Chats" broadcast nationally by radio, and resumed during the national energy crisis of the later 1970s, when President Jimmy Carter, informally dressed in a sweater, spoke to the nation in televised chats against the background of a crackling fire.

The chauvinism of the Colonial Revival, then, masks the historical anxiety just beneath the surface. The fairs, the copies of "colonial" furniture, the nostalgia and reverie for an idealized "open hearth" American past occurred during periods of social turbulence and distress. It has been remarked that those Americans participating in the Colonial Revival deeply wished to enter into the nation's history. The reverse, however, is actually the case. The embrace of the hearth was really a way out of the discordant, tumultuous processes of history and an entry into American myth. That myth says that social change is not occurring as long as the hearth remains in place. Political shiftings, violence, social hierarchy, transformations of family patterns and demographics—all these are denied by the symbolism of the hearth. Dynamic processes of virtually every kind are denied. In fact, the meaning of the hearth is that history does not exist.

The TV hearth continues this message, both in the era of the Cold War and now in the era of nuclear threat, terrorism, and Middle Eastern conflict. In 1947, at the close of World War II, Secretary of State Dean Acheson warned that "the nation must be on permanent alert" (May 90). In Elaine Tyler May's recent study of American families in the Cold War era, one chapter is aptly subtitled, "Cold War, Warm Hearth," and she writes that "Americans were well poised to embrace domesticity in the midst of the terrors of the nuclear age. "A home filled with children would create a feeling of warmth and security against the cold forces of disruption and alienation" (23). By 1959, May reports, two of three Americans cited the threat of nuclear war as the nation's most urgent problem, and the sociologist Todd Gitlin remarks of his youth in the 1950s–1960s, "The grimmest and least acknowledged underside of affluence was the Bomb . . . a menace but a dangling one, not only abstract but oddly impersonal" (*The Sixties*, 22, 318). Civil defense drills during the 1950s emphasized preparedness as the key to nuclear age survival. The home bomb shelter was widely publicized as a way to promote family security in a hostile and dangerous world, and mass circulation magazines featured articles on the planning and purchase of these basement shelters.

The hearth emerges in this formulation as television set. In 1960, *Life* magazine featured a "spare room" converted to family fallout shelter that was designed, at the request of the federal government, by the American Institute of Decorators, and presented in a cutaway model to onlookers who gaze through the cinderblock wall into a room intended to house an American family of five in the event of nuclear war. The shelter is filled with objects intended to reassure onlookers of normality and stability, for instance a child's rag doll and a coffee pot, and the exercise bicycle that suggests mobility, fitness, and even youthful insouciance. The global wall map indicates that the world outside is geographically intact and retains its political divisions, and the animals on the wallpaper (buffalo, horses) tell the public that the flora and fauna are growing and grazing in the world beyond the cinderblocks of their bunker (even if the animals resemble those of the much-publicized prehistoric cave drawings of Lascaux, France, and thus subconsciously, on the part of the interior designers, indicate that the family are cave dwellers in a world bombed back to the stone age after all).

And the television?—"battery-powered," says the caption, and in the *Life* photograph the eye moves to locate the television as the central focus of the scene. The ideological program of this entire domestic scene is radically disrupted if one asks the question, Who, in the event of nuclear war, would be broadcasting programs to our sheltered family of five? Governmental answers about civil defense "experts" sound like palliatives in a post-Cold War moment. But in its place at the visual center of the photograph, the television is more than a communications medium with the presumed outside world. In addition, it functions as hearth, bearing all the meanings codified from writers like Stowe through the Colonial Revival. It is patriotism and domestic security combined, and the inevitable gathering place for the American family in a hostile world. It is the fireplace in the cave.

This television hearth remains as the nuclear threat continues, if not

MODERN LIVING

Fallout shelter with television, 1960. *Life* Magazine.

between superpowers, then by a Third World nation or terrorist group. In a 1980s short story, "On for the Long Haul," by T. Coraghessan Boyle, a Los Angeles man, terrified that "civilization was on the brink of a catastrophe," buys Montana acreage from a real estate developer who specializes in bomb shelters and survival homes. He installs his family in a steel-plated "cabin" whose basement walls are four-feet-thick, lead-lined concrete. In the evenings, "when the house was quiet as the far side of the moon, [he] would slip down into the shelter, pull the airtight door closed behind him," and "at night there was television, the signals called down to earth from the heavens by means of the satellite dish" (Boyle, *Greasy Lake and Other Stories* 63, 69–72). And Walker Percy echoes nuclear threat amid natural cataclysm when, in his novel *Lancelot*, a hurricane bears down on a Louisiana community, and at last "owners of fallout shelters dug out during the A-bomb scare . . . and never used" became "happy families huddled underground around TV sets showing [hurricane] Marie spinning ever closer" (207).

Images of the electronic hearth in the era of imminent nuclear apocalypse are as explicit in recent fiction as in the *Life* photograph of the bomb shelter,

Pioneer Corporation ad, 1989.

and to scan contemporary narratives is to see the pervasiveness of the nuclear consciousness precisely as it makes itself felt in desultory moments in diverse texts. In Ann Beattie's *Love Always* (1985), a young woman living in Vermont in the summer notices white pansies blowing "like hankerchiefs held in the air," and associates them with "ladies waving to soldiers as the train pulled out," immediately realizing that such scenes had already passed into myth because "handkerchiefs were an anachronism, and the next war would blow the Earth away. Nobody was going anywhere by train" (15). Similarly, in Meg Wolitzer's *This Is Your Life* (1988), a teenage girl thinks about her boyfriend who seems to her intangible: "but it wasn't just her boyfriend who was absent. . . . It was *everyone* in the universe . . . as though there had been a nuclear holocaust, and everyone had been swallowed up" (64–65).

These kinds of moments indicate why the TV hearth is with us still, imaged once again in a TV ad, this one featuring the downsized family of the 1980s, a Yuppie family with one child living a life of excellent taste in minimalist style that rejects mere bric-a-brac in favor of a few well-chosen pieces (a soft leather couch, a designer telephone, a simple glass bowl holding low-calorie popcorn, and no movie-era butter to imperil health or smear the glass). And of course a big-screen television on which the family watches the movie, *E.T.*, imaged in the ad at the final moment when the endearing extraterrestial goes home. The family, of course, is already home, their lightweight clothing and bare feet indicating physical comfort and telling us they cuddle for emotional, not merely thermal, warmth. In the room, the worlds of television and the domicile are unified in patterns of light, the oval of on-screen luminescence repeated in the glowing, oval pools (doubtlessly cast from unseen track lighting above) on the exposed wall brick. Here the traditional material of hearth and fireplace—the brick—is present in the wall itself, which the ad asks us to contextualize as TV hearth in an era in which the cabinetry has receded in importance, the TV set no longer an important piece of furniture but a state of the art electronic appliance manufactured in East Asia (just as the family's sofa and telephone are probably imported from Europe). As for the rosy fire, it too is present. This black-and-white ad presumes our familiarity with the on-screen movie and asks us mentally to colorize the image of *E.T.*, who literally glows in happy moments throughout the movie, his chest the radiant red-orange of firelight at the hearth.

In a nuclear–terrorist era of internationalism, an era of the AIDs plague, food contamination scares, chronic racial hostilities, this image functions as a TV hearth in a continuation from the Cold War "hearth" of the bomb shelter. This family is, in late-1980s parlance, cocooned, bricked in, still sheltered against the fallout, be it nuclear, biological, or sociological. Their enclosed space indicates no source of natural light, no immediate access to an outside world. The child's pajamas, the mother's bare feet tell us they would not think of going outside. (Indications are, the parents will watch another, adult-rated movie, *The Untouchables*, when the boy goes to bed, or so the videodisk jacket on the rug indicates.) The reflected glow of the three faces is emphatically ethereal, but any spirituality seems compensatory for—or perhaps contingent upon—their walled-in, implicitly subterranean safety.

No outside world exists in this world principally of the dematerialized image. The ad reveals certain values in contemporary U.S. history, in which a nation that has lost its lead in the manufacture of objects manifests the dematerialized as that which is to be desired, from the thinline telephone and remote to the narrow-frame cabinetry and the chaste popcorn still, at the end of the movie, in supply in the thin glass bowl. Yet the brick and the glow tell us that the idea of the American hearth is still central, functioning as reassurance against a hostile world. The slogan of the Pioneer Corporation serves still to reinforce the hearthside values of patriotism and domesticity: "Pioneer—We bring the Revolution home."

And the TV hearth is still functioning as a time trip out of the tumult of history into the placid realm of American myth. No matter what discordant images may flash across the screen—wars, racial hatreds, disease, family violence, abject poverty, natural catastrophe—they are to be neutralized, in fact denied, by the TV hearth. The receiver itself reassures viewers that all images are as safely contained as the flickering firelight within the hearth. Such fires are never incendiary, never inflammatory precisely because the hearth domesticates them. The corporations, their sponsors, and the broadcasting industry have worked, as the Colonial Revivalists have long worked, to exploit the American hearth to promote domestic tranquillity.

Peep Show, Private Sector

When I woke up, snow fell softly at the window and the black and white television was on. Lucy and Ethel were trying to steal John Wayne's footprints from the cement outside of Grauman's Chinese Theatre in Old Hollywood.

I stayed in bed and watched the reruns and time fell off in half-hour segments. Then I got up to get something to eat.
Mona Simpson, *Anywhere But Here*, 1986

I spent the weekend alone in the apartment. There wasn't much to do—I got my clothes into the closet and then . . . bought a small color television, a Panasonic, which I watched all day Saturday and most of Sunday. When I wasn't watching television I was in bed, fully dressed, thinking. When that seemed self-indulgent I got up and watched more television.
Frederick Barthelme, *Second Marriage*, 1984

The TV was on. Always, wherever she was, a TV or a radio was going, a habit . . . of her many years alone.
Scott Turow, *Presumed Innocent*, 1987

A discursive mix of journalism, advertising, and fiction have promoted television as a modern hearth, but the scenes above imply radically different meanings. Each individual watching television in these scenes is alone. The hearth, so visual, so concrete and inclusive, seems like television's compleat cultural emblem—until we encounter the lone viewer trying to lessen his or her isolation by turning on television. Not one of the three fictional characters watching television in their bedroom in these scenes is gathered at the electronic hearth for an affirmation of sociofamilial domesticity or patriotism. Even in the 1980s–1990s, when it seems perfectly ordinary for individuals to have televisions in their bedrooms, each of these characters is emphatically by himself or herself, and each is disconnected from friends and family. They may turn to television for company (e.g., for the familiarity of *I Love Lucy*). But their *loneness* argues that the TV environment cannot be represented solely in the image of the electronic hearth, that a contrary, even contradictory formulation has developed over decades and demands disclosure.

The separate, solo TV experiences vividly show two clashing patterns, one communal, the other personally individual. Against the sociofamilial hearth, we have the challenge of decentralized television, a one-on-one experience in which television is a private and individualistic act. Not surprisingly, solo television claims the tradition of American individualism, with its emphasis on personal acts and choices. And its imagery, like that of the hearth, masks certain anxieties relevant to the very traditions it seeks to uphold. Individual, private preferences in TV viewing are constructed to reinforce long-term values of American individualism, which extends from the eighteenth-century Enlightenment and is a major part of the ideology of democratic America.

Individualism, however, is as problematic in the TV era as it has been in the past, because individual self-assertion can set the viewer apart from, and at odds with, the family, partner, or group, and so undermine the social symbolism of the electronic hearth. A fictional moment in Mona Simpson's *Anywhere But Here* (1986), which spans the mid-1960s–1979, shows this clearly. The protagonist, an adolescent named Ann, tells us that after school she enjoys lying alone on her stomach in the cool basement watching TV reruns—until her mother appears with laundry and, seeing her daughter alone watching television, grows suspicious and launches a series of accusations: "What are you doing down here? Why do you always come down here when you could watch the big TV upstairs? Are you ashamed of something?" (99).

The vignette suggests that individual privacy is itself an index of culpability. It reveals the tension between the social hearth and the private, individual televisual experience. Because both ideas of television, social and individual, claim the same persons, the same viewers, the electronic hearth and individualist television potentially rival one another. When positioned against the social hearth, individualist television can threaten the ideology of that hearth, contradicting it, betraying it and, at the same time, exposing private, individual television as selfish, egoistic, narcissistic, socially dispersive and, most disconcerting of all, subversive. Early articles about television had "showed a family cozily sitting together before the television set," wrote a critic in 1977, adding, "Who could have guessed that twenty or so years later Mom would be watching a drama in the kitchen, the kids would be looking at cartoons in their room, while Dad would be taking in the ball game in the living room" (Winn,

40). Cartoon humor even at its gentlest conveys this anxiety about social dispersal. In one drawing, a party guest, bored by home movies, slips away to switch on his host's television to see what he might find to amuse himself. In another, the housewife, fed up with her husband's monopoly of the television, orders a receiver of her own, delivered through the front door while the husband in the living room watches his baseball game. Both cartoon characters have taken private initiative in defiance of etiquette and domestic order (*The New Yorker*, 31 [32] [Sept. 24, 1955]: 44; 33 [30] [Sept. 17, 1949]: 33).

To understand what is at stake here, we must glance at individualism itself, long a hallmark of American identity, its exponent Ralph Waldo Emerson, whose essay "Self-Reliance" (148) had enjoined the white American man, "Trust thyself: every heart vibrates to that iron string." Emerson alleviated any potential problem of the separate American self or act by presuming that self to be white and male, and assuring him that everything that is "absolutely trustworthy" is "seated at their [meaning his] heart" (501). In the individualistic world of Emerson's America, the center is the heart itself, wherever one happens to be at any given moment (Bercovitch, Jehlen). In the 1950s, President Dwight D. Eisenhower reaffirmed this outlook, insisting that "more than ever before, in our country, this is the age of the individual" (Ward, 260).

Yet the individual continued to be a controversial figure in the TV age. Postwar educator–scholars, for instance, debated the nature of authentic and false American individualism, which is to say individualism assumed truly to exemplify democratic national values or contrarily to subvert it from a state of mind better defined as alienation. The topic continued to be vigorously debated in relation to social cohesiveness. The critic, John William Ward, for instance, discovered that individualism was redefined, no longer meaning *"freedom from society"* but *"freedom to cooperate"* in society (231). And the sociologist David Riesman, probing the relation between self and society in *Individualism Reconsidered* (1953), found that while lip service was paid to "rebelliousness" and to "difference," socially conformist terms like "team spirit" and "popularity" were particularly important, reinforced by schools and other institutions, by the workplace and the media. "Popularity" and being "one of the gang" were really synonyms for conformity and convention, Riesman argued, and President Eisenhower confirmed this very point in defining American individualism as a configuration of "free individuals joined in a team with our fellows" (Ward, 260). Such consensus terms functioned precisely to allay anxiety about individual singularity taking alienationist, antisocial forms. The ethos of "cooperative team spirit" masked evident anxiety especially about individual and private male thought from which the subversive, the un-American, or the antisocial action might erupt.

Yet the oxymoron of individualistic teamwork did not cover all contingencies, and the nature of individualism was argued with particular vigor in relation to conspicuous cultural figures, such as Herman Melville's Captain Ahab and the Henry David Thoreau of *Walden* (May, 162–63). Postwar critical consensus held that Melville's long-obscure novel, *Moby Dick*, was actually an important American epic, a national classic. Readers and commentators were thereby forced to confront the problem of the conspicuous individualist, Captain Ahab, the monomaniacal ship master who forsakes his family life, which is to say his hearth, and binds his crew to a state of subordination that leads to their destruction.

These same readers found themselves simultaneously confronting the individualistic Thoreau, who in the mid-nineteenth century absented himself from hearth and home in Concord, Massachusetts, to take up a solitary existence in a woodland cabin and to produce his account of that experience, *Walden*. Thoreau, as one current critic observes, "elevates disconnection into a national ideological value," and "etherealizes friendship to the point of mutual evanescence" leading to "the separation of individuals" (Pease, 262–63). Thoreau withdrew from society, while Ahab took it with him, bending it to his will and destroying it. Both represent extremes of American individualist action, and both were figures interpreted to the thousands of Cold War-era students studying American literature, which had "acquired legitimacy and prestige as a subject fit for scholarly study" even as television proliferated in the households of these same students' families. (Cain, 138).

In what terms does the vexed issue of American individualism enter the forum of television? President Eisenhower and the scholars notwithstanding, it can be argued that commercial television actually hastened the debasement of individualism from the late 1940s onward, turning it into mere marketplace consumerism. Emerson's aesthetic and philosophic principles can appear to be lamentably degraded in the TV era—Ralph Waldo yielding to the TV sets of the Emerson Corporation. Juxtaposed, the two Emersons, the philosopher and the corporate electronics manufacturer, make us aware that in the consumer age, individualist America in large part has meant the freedom to "project one's desires onto produced goods," to "actualize [the self] in consumption, each on his own," and that personal liberty has thus taken the form of individuals' acquisition of products and services, including TV receivers and programs (Baudrillard, 13, 12).

Certainly the discourse of manufacturers and marketers encourages this viewpoint. By the 1970s, the Sony Corporation ads for small-screen portables could exploit this individual preference—the lone soccer fan at the neighborhood tavern whose other patrons absorb themselves in baseball, the wife enjoying the morning program at the breakfast table while her husband reads the newspaper. The Sony ads point up the commercial drives for decentralized

The Sony for Soccer Fans

Sony Corporation ad, 1967.

television: the impetus to sell more receivers by creating consumer demand for multiple TV sets for bedrooms, kitchens, suburban garages, "family" rooms, and so on. Like the hearth, the decentralized TV environment focuses on the habitat. In 1951, the DuMont Corporation identified the "two-set home" with two photographs, one atop the other to suggest the upstairs–downstairs scene at evening, when the parents watch the ballet in the living room, while below, in the "rec" room, their child(ren) and friends watch a baseball game. "Think of the advantages enjoyed by the two-set home," says the ad copy, "—freedom for young and old to enjoy favorite programs at any time." The console or floor models will "grace" the living room, while "compact table models" can be dispersed in the "den, library, or game room."

The "compact table models" are just one step toward portables, like the 1960s Sony "Tummy Television," or the RCA "Sportabouts" advertised with a photograph of three men, one in a business suit, two in sportswear suggesting boating and, perhaps, bowling. Each has chosen a different model Sportabout ("The Entertainer," "The Wayfarer," "The Roommate") and so affirmed his individuality. There is emphasis on individual selection (one television per person, a separate model for every individual and, implicitly, a program for each individual taste).

Potentially, however, the proliferation of portable TV receivers poses an ideological threat to the hearth. Individualistic program preference could lead to the dissolution of the family center. Should members of the habitat disperse into separate spaces, each in a self-enclosed relation to the on-screen world, then the very idea of familial "togetherness" would become a mockery.

Marketing, however, diffused the threat by redesigning the TV receiver and by ad copy that configured the viewer as a vacationer or tourist. By 1960, the casings of portable TV receivers were shaped like luggage with handles on top, suggesting a mere sojourn or vacation, and the very names of these televisions were anodynes to a buying public encouraged to think of the personal television as television only *temporarily* in private use. The suffix, "about," in the RCA "Sportabout" and "Tote-about," for instance, emphasizes the sense of a circuit traveled all around from a center. The viewer became a sojourner or vacationer, carrying the new television like a suitcase. The televisions are seen in their state of portability, the men walking with them photographed while in transit.

But the center—the hearth—is presumed to hold fast. In fact, the electronic hearth makes the luggagelike portable television possible. The ad says that these consumers, notwithstanding their individual selections, are only absenting themselves from the hearth for a little while, for the duration of a game or a cherished program (or, in the case of the "Roommate" model, until they establish a hearth of their own). In every instance, the individual *cum* portable

Everybody

enjoys television in the two-set home
...and enjoys it more with *Du Mont*

Think of the advantages enjoyed by the two-set TV home — freedom for young and old
to enjoy favorite programs at any time. And for your next set, look first to Du Mont.
Choose from magnificent consoles to grace your living room, or compact table models
for den, library, or game room. Each is a masterpiece — built with precise circuitry,
over-size parts, extra parts — to give you the finest possible performance.

DuMont Corporation ad for the "Two-Set Home," 1951.

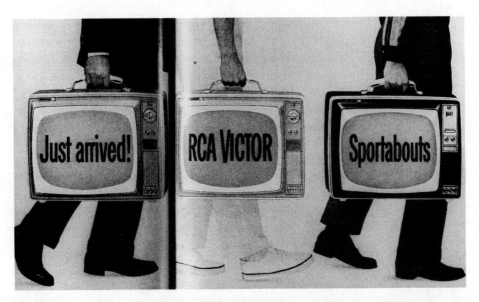

RCA ad for portable television, 1960.

General Electric ad for portable TV "luggage," 1965.

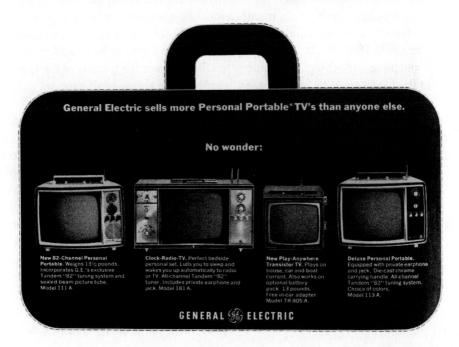

is removed from friends and family only for an interlude. Similarly, Sony's mid-1960s "Tummy Television" functioned linguistically to assure the viewer (and those excluded from the on-screen view) that the TV receiver is positioned so exclusively, so privately for the benefit of one individual *for just a little while.* The "tummy," after all, can only support the receiver when one is supine. Even if the "portable" becomes a permanent fixture in the antipodes of the habitat, luring an individual habitually away from others and to it, its name and the semiotics of its design continuously allay anxieties about privacy or defection—or disaffection—from others. It is *only* a *portable.* The individual using it is only taking a short time-out from the home-hearth.

The marketing of multiple televisions for each habitat shows how manufacturers and ad workers have walked an ideological tightrope between human centrality and dispersal. They have been careful to valorize the electronic American hearth even as they promoted the multiplicity of televisions for individuals in every niche of the dwelling. For the decentralized TV environment, promulgating dispersal, always risks betraying the hearth by exposing tensions and divisions between generations and between sexes. In cartoon humor, not surprisingly, decentered television features marital discord and family tensions. One cartoon of 1956 suggests this very tension by spoofing its resolution. The cartoon shows a family of three gathered before three lined-up televisions, mother, father, and son happily watching different programs while the audio comes through headphone sets that each wears. Gathered together as a nuclear family, each member inhabits a separate on-screen world. By extension, a second child would mean an additional receiver and headphone set, and so on. The family's contentment comes from *not* having to gather to watch the same thing. Technology lets them escape the tyranny of the hearth (*The New Yorker,* 32 [30] [Sept. 15, 1956]: 38; see also 24 [39] [Nov. 20, 1948]: 48).

Yet evidence suggests that the acculturation of individualistic television involves the suppression of other anxieties directly relevant to individualism in a protracted Cold War era. We can gain entry into these masked or suppressed areas of cultural angst through an unusually confessional essay, entitled "What We Do in the Dark" (orig. publ. 1976), in which Michael J. Arlen, the former *New York* TV reviewer, makes his own household a case study of the experience of decentered television. "I happened to pass by the little side room where we keep our TV set," says Arlen, and "spied the sprawled, glazed-eyed figure of one of our children . . . bathed in the gray, flickering darkness." Arlen admits that "there is something about him [Arlen's son] in that darkened TV room that irritates me beyond measure," and he goes on to say that his wife, also passing by, caught a look at Arlen's face and asked, What was going on in the TV room? "You looked," she said, "as if you'd just come across somebody 'in the act.'" Arlen denies this but asks himself why he "vaguely patrols the TV room without acknowledging that he's patrolling it," and goes on to describe "the electricity that seems to surround this apparently so ordinary, domestic activity, television viewing" (202–203).

In Arlen's analysis of his typically middle-class American domicile, what is really going on in that little side room is sexual excitation. He identifies the unstated "guilty recognition of the sexual connotations of watching television" and acknowledges that while TV watching is considered a group activity, "most of us engage in it primarily on our own" (201). Often, he says, "we set up a special, private room at home to do it in, for above all it is a private act, an act that we perform by ourselves and with ourselves." Sometimes, "we try to talk about what we do in our dimly lit, flickering rooms," he says, "but it always comes out that we were . . . watching a show, a football game, our 'favorite program.'" Instead, says Arlen, what TV watching most resembles "is masturbation" (204, 201).

Masturbation. Here Arlen identifies what he believes to be the televisual secret of middle-class America, that there is no communal intimacy of the hearth, that TV watching is private and autoerotic and that unacknowledged guilt and suspicion surround it. The private act, in his view, becomes the autoerotic act. Decentered television is the only kind there is, and the privacy of it becomes masturbatory.

The graphics on decentralized TV watching support Arlen's position. They are conspicuously gender-specific (and blatantly sexist and sexual). Cartoons, for instance, lampoon the housewife for neglecting domestic chores while watching daytime television romance—but assure us that this same housewife will not stray from the domestic sphere. Distracted by a couple kissing on-screen (presumably in a soap opera), a housewife in one cartoon forgets the iron and scorches her husband's trousers, oblivious of the column of smoke rising from the ironing board. Or she sits before the television, enervated amid domestic squalor, while on-screen the superhero Mr. Clean shows his biceps and the commercial jingle plays. Or, in a series of cartoon panels, she spends her day socializing, reading, and watching television (again, romance as the on-screen couple lean to kiss), only taking up the apron and vacuum cleaner as theatrical props of drudgery when her husband is due home from the office (*The New Yorker*, 36 [40] [Nov. 19, 1960]: 55; 26 [15] [June 3, 1950]: 22).

In ads, the sexual emphasis continues. Seeming to watch television from the bathtub, the woman in an Emerson Corporation ad, presented in a line drawing by the artist Saul Steinberg, is actually watched by an on-screen, three-dimensional man who adjusts his glasses for a better frontal view. Her back brush is really his right arm reaching figuratively around in a possessive caress. The scene stops short of lasciviousness only because of the man's

Housewife distracted from ironing by couple kissing on television, presumably in a soap opera. Courtesy *The New Yorker*. Reprinted by permission.

stereotypical intellectualism, announced in his opened hardbound book and his gaze of studied connoisseurship. Such images, in ads and cartoon humor, underscore the message of erotic romance, but more powerfully emphasize the idea of domestic containment. In the bath, the television-era woman is monitored, under surveillance, and even when television distracts her from household duties by absorbing her in vicarious romance, she remains in the habitat where she belongs. Television, which she watches by herself, alone, may distract her with erotic fantasy, but it also contains her. It watches her, and she it, binding her to the household. Sexist images essentially show the containment of individual actions.

The male version of individualistic television is more emphatically sexual, configured in erotic heterosexual fantasy but, again, cast within a world of domesticity. On moving day in the suburbs, the husband (and the moving men) in a Norman Rockwell-like Motorola ad enjoys the on-screen image of a buxom female trapeze performer showing her decolletage while his wife, pausing with her broom, looks askance at the scene. Another husband (in an Emerson ad with a drawing by William Steig), lies beside his wife in their double bed, flirting with a woman singer on-screen. She watches him, and he returns her gaze, while his wife, aware of the flirtation, fingers the TV remote button, ready to disconnect her husband from his fantasy.

Versions of this male erotic heterosexual fantasy persist in images of the luscious TV temptress, such as the Sylvania/GTE color set Eve with her apple, or the Admiral Corporation "Playmate" in the decade of Hugh Hefner's *Playboy* clubs, or in the numerous line drawings and photographs of ads and cartoons showing the men in midlife yearning at bedtime, in bed, for that playmate. The images say that the American man, watching television by himself, is girl watching, enjoying sexual fantasy. Any potential threat posed

Emerson Corporation ad, 1965.

Wherever you look ... there's Emerson

Emerson Corporation ad, 1956.

by these private fantasies is diffused by the encompassing context of domesticity. In the ads and the cartoons, the man tends to be a midlife figure, rendered "safe" in his very physiognomy, his thinning hair and paunch telling us he will not abandon the domicile no matter how compelling the erotic fantasy. The graphics of TV erotics really provide assurance that decentered television is socially conventional, that is, natural, and therefore socially safe.

And fiction, too, makes this point, casting the moment of televisual fantasy as linear narrative. Frederick Barthelme's novel, *Second Marriage* (1984), shows the suburban husband–narrator watching television in an apartment where he lives temporarily upon the breakup of his marriage:

The television was still on [when I woke up], but the screen was snowy. . . . . I crossed the room and twisted through the channels. . . . The tireless people of the Christian Broadcasting Network were still at it. A shy-looking girl with a broad face and freckles came out from behind a curtain and sat on a barstool at the center of a bare stage, her legs crossed. . . . When she started singing I watched her left hand move over the strings. Her nails . . . glistened under the lights. . . . Everything she did seemed sexy. . . . I imagined her in [my] dream, sitting at the beach . . . the dress lifted, the clear skin of her thighs, the smile. I imagined loving her, kissing the skin, kissing between her legs . . . her clitoris was . . . something tiny and delicate and wet . . . sweet to the taste. I imagined lingering there.

Then her song was over and the announcer . . . came into the picture and started talking to her. . . . I snapped off the television. (188–89)

This is one of two extended moments in the novel in which television is the occasion for sexual fantasy, and the domestic habitat its location. Here the novelist can narrate the course of the fantasy-as-narrative, and suggest additionally that television itself structures the fantasy, that its change of scene

American Airlines ad, 1960s.

cobean

truncates the erotic moment, that the male viewer is held in check by the very pace of the television's short segments. If the Christian network can incite sex fantasies, television itself controls them, and the habitat (here, the pastoral suburb) contains them.

The sexual stereotypes of men and women in these images proclaim that television is an erotic environment for women and for men, yet provide assurance that the firm context of domesticity will restrain individualist sexual energy, confine it to intermittent moments of vicarious pleasure. Individualist TV, then, seems to divide along rather predictable lines susceptible to easy interpretation. The sexist politics of these texts are exactly what feminism has found objectionable from the late 1960s to this day. And the containment of women and men in these stereotypical poses of fantasy is remarkably similar to the containment implicit in Arlen's "What We Do in the Dark," in which parents and children cohabit in the context of the nuclear family, the father patrolling the apartment halls, the mother monitoring the scene, the son–viewer under scrutiny there in the darkened TV room.

But beneath the issues of sexuality and television lie deeper anxieties about the decentered, individual experience itself, and Arlen participates in these anxieties in revealing ways. His analysis leads into implications concealed by the sexual politics of the ads and cartoons, and for that matter by his own essay, which tells us that privacy is integral with autoerotics in the television environment. Arlen's conjunction, similar to that of the ad and cartoon graphics, is itself revealing. Televisual privacy indicates sex. The two, sex and privacy, are conjoined, with suspicion and guilt assigned as effects, spinoffs. "What We Do in the Dark" lets readers watch the way in which Arlen, like the cartoons and ad graphics, affirms the conjunction of sex with privacy. It is sex that explains privacy, in effect normalizing it, configuring the guilt and suspicion as a part of traditional, middle-class mores. It may be jarring to hear TV watching described as masturbation, but it is not deeply disturbing. It is, on the contrary, finally reassuring. "Is *this* what people do in the dark? Is *this* what we do, too?" Arlen concludes his essay (204). Once past the initial indignation, the reader might add, relieved, "Is this *all* we do in the dark"?

This is not to redirect Arlen's essay but to notice how his conjunction of sex and of private acts functions to displace the very idea of the private. In Arlen's essay, as in the graphics, televisual privacy is by definition autoerotic; engagement in the TV environment is per se autoerotic. All other areas of thought and feeling are precluded. And this, of course, is also the message of the cartoon humor and of the advertisements for decentralized TV—romance for the ladies, sexual fantasy for the men, and both so traditional, so natural that there need be no anxiety aroused on other grounds.

TEMPTING

NEW COLOR TV BY SYLVANIA

General Telephone and Electronics ad, 1960s.

This very message, then, works to contain, to manage the experience of individualist television by precluding other thoughts and feelings.

What might they be? The novelist Don DeLillo sketches two possibilities. In his novel, *Libra* (1988), the boy Lee Harvey Oswald sometimes sits alone at night watching TV crime shows (*Racket Squad, Dragnet*) or tuning out his mother by staring at the on-screen DuMont test pattern (*Libra*, 5–6). Far from erotic fantasy, private TV watching is here an incubator of presidential assassination. And DeLillo's *White Noise* (1985) reinforces the pathology of

Tummy Television

The 5 inch Sony, for waist sizes 38 to 46. (For smaller tummies, buy the 4 inch set.) Our 32 non-heating, long-living transistors plus our telescopic antenna give you flicker-free reception—even if you jiggle when you laugh. The Sony works on AC wall plug or clip-on battery pack. So that your wife can sleep, we also include a personal ear plug. The beauty of a TV set this small: when you've had a bellyful of television, you hide it under the pillow.

Lightweight 5 inch SONY TV

Sony Corporation ad, 1965.

individualist TV when we learn that a rooftop sniper had been hearing voices on TV, "insistent pressuring voices . . . telling him to go down in history" (44–45). It is television, DeLillo argues, that instigated Jack Ruby's murder of Lee Harvey Oswald—"TV gave the clue and Lee was shot. . . . This boy was shot handcuffed to an officer of the law. . . . TV gave directions and down he went" (*Libra*, 450–52). The novelist affirms the act of violence perpetrated by the isolated individual in response to television directives. DeLillo describes the very pattern found in film and contemporary history, in which William Hinckley, influenced by the Martin Scorsese film, *Taxi Driver*, and obsessively drawn to the actress, Jodie Foster, who performed in it, attempted in 1980 to assassinate President Ronald Reagan. The novelist or film maker, of course, has latitude foreclosed to the advertising worker, who could not seriously propose marketing television (or movies) as a training ground for killers. Nor does Michael Arlen even hint that decentered television promotes sociopathic tendencies. These recent fictional examples, however, crack open the sealed world of erotic, private television. They show the extent to which the sexist images in advertising and in cartoon humor have managed and contained the greater anxiety about private TV watching—especially men's—as an incentive to dangerous acts of renegade individualism.

The crux here is the individual male in isolation, potentially acting alone in ways that pose a threat to the very TV-age culture that has proclaimed togetherness and teamwork to be cardinal virtues, yet trailed the Emersonian ethos of self-reliant individualism. The TV viewer can be observed in the act, but the act itself is private, concealed, and evidently much more threatening than, say, reading, which has been enshrined as an intrinsically valuable activity equated with education and self-improvement. Even if one is not reading "great" literature but "junk," the act of reading per se signifies literacy and self-discipline, and the provisional possibility that the reader, once surfeited with banal texts, might move on—move up—to the great or the greater books.

Non-TV acts of private visual engagement, at the same time, tend to be culturally circumscribed and officially authorized. In U.S. cities, the Red Light districts with their peep shows and pornographic bookstores have been geographically zoned by politicians and newspapers, and patrolled by police whose periodic raids reinforce the idea of surveillance and of sociopolitical containment. As for the experience of the visual in high culture, expositions and exhibitions have been arranged by arbiters of taste and authority (e.g., museums, corporations), as are theatrical performances, dance, and the like. The unchallenged assumption is that the visual engagement is personally edifying to the onlooker. The individual benefits educationally, culturally (in high cultural terms of edification and improvement), intellectually, and aesthetically.

But television, like the movies, is an unbounded visual experience, controlled by advertisers and networks and the FCC (and in the case of films, the Motion Picture Association ratings), but unzoned, so that the very threat of autonomous private acts, fostered by solo watching, lies just under the surface. That threat is important in the TV age because it lies at an unacknowledged juncture between two concepts of the self—the first individualism and the other alienation. The vignettes of Lee Harvey Oswald, Jack Ruby, and the rooftop sniper play an important part in the image of the single male television viewer because they argue that the individual buying a portable television to watch alone is potentially much more than a debased Emersonian driven by modern consumerism. The vignettes imply that the tradition of individualism cannot be so dismissively relegated to the commercial sphere. Individualism is necessary to the image and self-image of the American man shown to be by himself before the TV screen, because it is the ideological concept that explains and permits masculine loneness in a secular realm, including that of television.

Individualism, in fact, is the protective coloration against DeLillo's Lee Harvey Oswald or any other homicidal TV watcher. As the novelist suggests, the TV watcher already deeply isolated from society is a figure whose individualism veers into alienation. Historically, as one scholar writes, "millions are said to have been suffering . . . from social alienation and to have expressed it by their indifference or hostility toward most of their fellow men or toward society's organization, workings, and goals" (Feuerlicht, 76). Disaffected, the alienated individuals become potentially threatening. The individualist's solitude becomes the isolation of the alienated man. The TV voices heard while watching all alone can become incentives to hostile, dangerous acts. And how is the benign individualist to be distinguished from the dangerous alienationist? Most discussions of American individualism avoid that hard question by shunning the very connection. Studies of individualism and of alienation sit for the most part on separate library shelves as if to preserve a taboo against entangling the two concepts (e.g., Gans and Ward, Feuerlicht and Kaplan).

Yet we need to appose them here. The TV-age Emersonian individualist benevolently centering the universe has a doppelganger in the alienated individual who sees no meaningful connection between himself and his social relationships, his status, his work, his style of life, or any of his identifications (Kaplan, 118). The alienated self, the Oswald or Hinckley, may see social rules and governance as barriers to autonomy and violate them to attain true "freedom." As Morton Kaplan remarks, the alienated man "is continually driven to further extremes [because] the world he seeks to unite to him eludes him" (139). And one recent study of individualism in the United States, *Habits of the Heart* (1985), acknowledges in brief that "sometimes the flight from society is simply mad and ends in general disaster," instancing *Moby Dick*'s Captain Ahab as the consummate example of "asocial individualism" in America (Bellah et al., 144–45).

In fact, throughout the protracted Cold War era of "togetherness," an era of domestic sitcoms like "Father Knows Best," "Ozzie and Harriet," "Leave It to Beaver," "The Partridge Family," "The Brady Bunch," "All in the Family," and "The Cosby Show," the figure of Ahab has been a recurrent touchstone of American individualism run amok. In an era when individuals with portable televisions were imaged as temporary sojourners from the hearth or as domesticated men indulging in heterosexual fantasy prompted by the television image, the figure of Ahab has been a monstrous American constant in cultural criticism. Very recently, a critic has cited him as an exponent of American imperialism bent on ravaging the natural world with the aid of a colonized Third World crew (Dimock). But Cold War critics have seen him differently, not surprisingly as an exemplar of socially destructive individualism, the "embodiment of his author's most profound response to the problem of the individual free will *in extremis* . . . he can see nothing but his own burning thoughts since he no longer shares in any normal fellow-feelings . . . [and] refuses to be deflected from his pursuit by the stirring of any sympathy for others" (Matthiessen, 447–51). The Cold War's Ahab is thus a "fearful example of the self-enclosed individualism that . . . brings disaster both upon itself and upon the group of which it is a part" (459). He is "a false culture–hero, pursuing a private grievance (rather than a divine behest) at the expense of the mankind in his crew . . . a Satan, a sorcerer, an Antichrist" (Hoffman, 234). In these terms *Moby Dick* becomes "a book about the alienation from life that results from an excessive or neurotic self-dependence," one in which Ahab is "guilty of or victimized by a distorted 'self-reliance'" (Lewis, 105). The spokesman who celebrates *Moby Dick* for embodying the "great cultural heritage" of the United States warns nonetheless that Melville, the author, must not be seen to approve Ahab's "intensity, power, and defiant spirit" because he represents the deformation of individualism and self-reliance (Howard, 176–77).

Not coincidentally, the American movie-going public had a technicolor version of this same Ahab in the 1956 Warner Brothers film featuring Gregory Peck as Ahab. The movie, regularly aired on television and available for video rental, claims the values of domestic order and of capitalist business, with several shots of the women (wives, sisters, mothers, widows present and future) left behind as their men ship out, doomed by Ahab's vengeful quest. At sea, the courageous first mate, Starbuck, played by Leo Genn, voices the values of domesticity ("If Ahab has his way, neither he nor me will ever see home again"), and validates risk-taking only in the name of entrepreneurship within the capitalist economic system and its laws. Whaling, Starbuck says, is "a service to mankind that pleases almighty God," and in vain he begs his fellow officers to "join with me and obey the law," rejecting Ahab's illegal blasphemy, his mission of hatred that moves the captain to act, as he himself acknowledges, "against all human lovings and longings," including refusal to join a neighborhood search, so to speak, for a twelve-year-old boy, the son of a fellow New Bedford sea captain lost overboard in the same whaling waters Ahab sails. All these movie events occur in the novel, and the Ray Bradbury–John Houston screenplay includes a pastiche of Melville's prose. But it is instructive to see that the film version mirrors the ideology of the contemporary educational public discourse on Ahab. He is America's worst nightmare about renegade individualism, which is really alienation.

And if the American, Cold War version of Captain Ahab echoes the horrors of Hitler and totalitarianism (and retrospectively, in U.S. history, indicts the unchecked predations of the robber barons, as F. O. Matthiessen remarked), he also stands as a warning against the ultimate threat of the decentered individual. Ahab, the renegade individualist, not only repudiates the hearth but destroys those committed to it. He must be condemned on sociofamilial grounds. He is the alienated individual who, uncontained and unbounded, must be ostracized.

And just as Ahab was the spectre of American alienation, another cultural figure emerged as his opposite and became the exemplary individualist. Henry David Thoreau, who withdrew from society to live alone in a cabin by Walden pond has been presented in the Cold War, televisual era, not as the advocate of an ideology of human disconnection, but as the exemplar of the democratic common man, recommitting himself to American traditions of hard work and craftsmanship, his *Walden* also making a claim for "communal security and permanence," for "order and balance" (Matthiessen, 172–73). He is a "visionary hero" who "demonstrates his freedom in the liberation of others," thus working on behalf of his fellow human beings to further the greatest cause of freedom (Lewis, 21). Thoreau's "aggressiveness," though regrettably "excessive" for his time, is vindicated when his democratic "symbol of the hermitage," is exported to India for Mahatma Gandhi's successful twentieth-century struggle for democracy (Howard, 158–60).

Thoreau, then, becomes the democratic heroic individualist precisely because he represents the dutiful and responsible American working man committed to communal security and stability, and also to order—in other words, he is a version of the very man imaged repeatedly in individualist TV cartoons and advertisements of the white middle-class man, briefly enjoying his private TV fantasies but otherwise committed to domestic and community order and to security and the work ethic. The individualist in the TV age, then, must either be brought within the democratic fold, as was Thoreau, or, failing that, consigned to unAmerican realms beyond, that is, to the alienationist realms of insanity, abnormality, and deformity.

So has it been for the lone male TV watcher. He must not be shown in public discourse as a watcher of violent programs, not even of news programming with on-screen victims of war or terrorism. The lone male is not represented as watching gore or mutilation, lest the over-his-shoulder onlooker infer his secret thought as actively violent. He must not be represented as a potential mass murderer or a Lee Harvey Oswald but a law-abiding citizen and family member. Television-age images of the private male act of TV viewing have been, therefore, tightly confined to contexts of the domestic, the communal, the social. And TV receivers intended for individual use have been marketed as "Tummy Television" sold in receiver casings that look like luggage. The treatment of Ahab by the spokesmen for intellectual America suggests to us why it is that images of men watching television alone, enjoying their intermittent heterosexual, even masturbatory fantasies, stand as reassurance and as behavioral examples that the self-enclosed, decentered viewer is not an alienated figure of latent violence precipitated by television. He will not become a murderous renegade individualist. He is the most domesticated of American men. What these images are really about is public safety in the private sector.

Leisure, Labor, and the La-Z-Boy

Leisure is the purpose of work. . . . The quality of leisure activity sets the tone of any society, defines its version of the Good Life and measures the level of its civilization.

Life Magazine, December 28, 1958

A resounding chorus of voices now proclaims that we are headed toward an unprecedented era of leisure in American life.

Robert Lee, *Religion and Leisure in America*, 1964

The basic point is that television must be viewed as a passive, noncreative, uninvolving kind of experience, and one which would not saturate the leisure hours of so many . . . if attractive and recreational opportunities were readily available.

Richard Kraus, *Recreation Today*, 1966

From midcentury, the figure of the TV watcher threatened to subvert one of postwar America's most idealistic and intellectually compelling visions of the immediate future. The consumerist viewing public was asked to see itself as family members clustered around an electronic hearth or as individuals sojourning with their portables, but the discourse of leading social theorists disrupted both formulations. The theorists, foreseeing automated factories as the engine driving unprecedented production and prosperity, hailed the imminent era of leisure in utopian terms—though their utopia was marred by one image, that of the TV watcher. As long as that figure was kept out of the picture, their statements could ring with utopian fervor legitimated by their scholarly and intellectual credentials. Representing social and cultural institutions, the theorists could speak as wise prognosticators of a future already taking shape. Omitting the figure of the TV viewer, these social theorists could advance their argument that the activist American work ethic should culminate in industrial automation, itself the gateway to a perpetual leisure society in the United States.

Once the TV viewer entered that projection, however, the optimism vanished and the vision turned to crisis, especially when the viewer was seen as an adult man settled back in his Stratolounger, BarcaLounger, or La-Z-Boy. That figure, shown in advertisements for consumer products, was recurrently represented as an enthroned monarch surrounded by abundant material possessions and a quiescent family. But when that figure enters the world of the social theorists' vision, leisure becomes passivity, and the work ethic mere sloth. Utopia, in short, deteriorates into an electronic-age Lotus Land. From the academy and the foundations, from sociology through journalism and corporate ads, TV discourses enacted a representational crisis both divulging and deflecting this crisis of leisure, labor, and the La-Z-Boy.

"Leisure Could Mean A Better Civilization," reads the opening editorial headline of a December, 1958, *Life* special issue entitled "The Good Life." The opening pages of the magazine featured a profusion of illustrations showing prominent Americans, from movie stars to corporate executives, engaged in leisure activities—sailing, model railroading, gardening, tinkering with automobile engines, cooking, biking, woodworking, arranging flowers. There were nineteen men and six women, including one African-American of each gender. Most were quoted on their understanding of the meaning of leisure. The physicist James Van Allen said that woodworking freed his mind, while George Romney, the head of American Motors, reported that basketball helped to keep "the body, the temple of the spirit, healthy and clean," and the then-wife of Nelson Rockefeller proclaimed gardening to be "a passion" she couldn't "wait to have time for" (78–79).

Nowhere, however, in this tutorial on leisure, targeted at the six and one-half million readers of *Life*, was there mention of television. The leisure hours of every prominent figure who was interviewed were reportedly spent entirely without it. In fact, the very existence of television in Americans' leisure hours was acknowledged only contemptuously and in patronizing tones throughout the special issue, for instance in an essay arguing that the millions of people reportedly "glued" to their TV sets were either children or else "adults who through personal misfortune have seen the great opportunities of the world shrink down . . . to the size of the tiny screen" (121). The unfortunates were similarly disparaged by a humorist who divided the American "new leisure class" into four groups from "Aristocrats" to "Peasants" (85–89). These last, the peasants, experts "only at wasting time," were shown spending their leisure in outré hobbies like paint-by-numbers, shooting jackrabbits from car windows—and watching television. Otherwise, though *Life* regularly carried full-page advertisements for TV sets and had done so since the late 1940s, the editors of this issue seemed conspicuously to avoid linking TV to leisure, as if in combination the two became anathema.

A new American postwar leisure society seemed just in the offing to mid-twentieth-century social analysts, but as the magazine editors implied, the vision was clouded by the graphic image of that leisure "saturated" by TV watching. As the projected height of industrial sophistication, an America of automated factories affording the good life of material abundance and spiritual well-being to the majority of citizens was threatened by an image that made a mockery of that future. Although television and the leisure society might seem like two propositions at least potentially consonant with each other, in juxtaposition they became inimical. As the *Life* essays indicate, the image of the sedentary TV watcher was understood to be a graphic rebuke to the very idea of an America free of toil and want, an affront to the attainment, as *Life* projected it, of the classic ideal of a higher level of civilization. Juxtaposed, television and leisure mocked the idea of national energy and trivialized national achievement.

The mass marketing of TV receivers in the United States coincided historically with social theorists' and futurists' projections of a leisure society imminent in the postwar industrial period. As the latest in a long American tradition of such projections (see Segel), students of labor in the 1950s–1960s were predicting the shortening of the work week in statements ringing with quantitative exuberance: "The working man of a century ago spent some seventy hours on the job," while "he spends some forty hours per week at work and can expect to live about seventy years. This adds something like twenty-two more *years of leisure* to his life" (Lee, 37). The most respected voices from the academy and other intellectual circles argued that in contemporary America, work itself was on the wane, leisure in the ascendancy. David Riesman's influential *The Lonely Crowd* (1950) stated that American society had undergone "a shift from an age of production to an age of consumption," and that the era of hands-on craftsmanship had yielded to one in which bureaucratic structures predominated in a corporate culture whose means of production was the automated factory (6). Riesman, like other sociologists, regretted the mind-numbing repetition of factory work but anticipated the total automation that would soon eliminate these stultifying jobs and inaugurate an age of leisure (262).

Riesman had an acknowledged industrial-era antecedent in the American utopian vision set forth in 1888 in Edward Bellamy's best-selling *Looking Backward*, which portrayed a totally automated American industrial state in which the young, in service to the nation, undertook the most distasteful work for a finite number of years before moving on to intellectually fulfilling activities and, in their midforties, retirement from national service. In Bel-

lamy's vision, the retirees, still in their prime, cultivated the arts and philosophy in their leisure in a classic ideal that, to late-twentieth-century readers,
seems inadvertently insipid on that utopian's part. But the portrayal evidently
encouraged Riesman, or at least abated his anxiety about social change. Based
on current trends, Riesman predicted automated offices, factories, and supermarkets, foreseeing the imminence of a leisure society.

Other voices did become a chorus. In that same year, 1950, Norbert
Wiener, in *The Human Use of Human Beings*, asserted that by 1960 or 1970
factory labor would be entirely obsolete (185–8), and in 1964, Herbert
Marcuse, in *The One-Dimensional Man*, predicted that individuals' "private
and social existence" would transpire in the realm of leisure (37; see Halprin,
pp. 42–91). And C. Wright Mills agreed with Riesman that "the old middle
class work ethic—the gospel of work—has been replaced . . . by a leisure
ethic" (236). Despite radical differences in political outlook, all these theorists
concurred on one point, namely, that the United States was entering into an
era of leisure for all classes.

As for the sometime synonym of leisure, recreation, the *Life* editorial
carefully distinguished between the two terms. "Recreation," it said, "is any
kind of mental or physical change from work that enables you to work better,
such as a stenographer's coffee break. . . . More and better work is the
purpose of recreation and play, so leisure is the purpose of work" [62]. That
logic makes recreation a stepping stone to leisure, and the recreationist press
joined in the intended spirit of the syllogism—yet did not include television in
its programs for increased leisure in the United States, calling instead for non-
TV family activities of the kind usually called wholesome. The journal,
Recreation, listed types of family recreation, among them talent nights and
amateur shows, bowling and skating, scavenger hunts, beach parties, county
fairs, hobby shops, flower arranging, and star gazing ("Family Recreation,"
340). As in the *Life* special issue, television was conspicuously absent, except
insofar as recreationists hoped it might move viewers to relinquish their
spectatorship and become recreationally active. "Can television . . .
change behavior patterns and convert youths and adults from spectators into
players?" asked a commentator who concluded gloomily that most viewers
unfortunately "remain spectators and not participants" ("TV and the Recreation Program").

Some theorists of recreation and of leisure sounded alarms directly about
television, fearing it could not be excluded or successfully resisted. In 1958, in
Reflections on America, the French social analyst and philosopher, Jacques
Maritain, wrote that, given automation and its accompanying shorter work
week, "leisure constitutes a serious problem for American life" because of the
problems of having "leisure time occupied in a manner really profitable to

man, and not entirely taken up by the sort of stupefying passivity that is more often than not developed by . . . television" (29–30). In 1954, a commentator wrote that "it is a fact that the average family with a TV set spends more time watching that screen in the living room than it devotes to *any other leisure-time activity*" (Wilson, 290). "Where will the breadwinner spend his (or her) spare time? Will our nation become a sedentary one, passively occupying television stools for hours on end?" (Floyd, 269). Here the term, "stools," refers to the bars and taverns where many U.S. citizens first encountered television, and rings especially pejoratively.

Thus intellectuals worried both explicitly and indirectly about the quality of this imminent leisure and the role of television in it. Their critiques often took the form of the idealization of a past Golden Age, whether of classical antiquity or an unspecified moment of pre-television, middle-class America. C. Wright Mills, echoing T. S. Eliot, deplored "the amusement of hollow people" who are neither calmed nor relaxed in a previous time of "old middle-class frolics and jollification." Instead, their leisure consists of "passive enjoyment and thrills" (238). And the radical Paul Goodman argued the contemporary impossibility of the classical ideal in which "serious leisure was the chief way that a free man grew in character and a city in culture." Instead, Goodman complained, the later twentieth century social theorists were speaking about neither culture nor civic duty but of "how to save the millions of people with long hours heavy on their hands from being drowned in canned entertainment and spectatoritis" (31).

Even the most roseate rhetoric boastful of industrial progress failed to extend its hymns to television, except insofar as the receiver itself signified purchasing power and consumer abundance. At base, visions of shortened work weeks with expanding leisure hours were not shown to be compatible with images of the individuals enjoying their leisure hours in front of the screen. As we have seen, the image of the family gathered at the televisual electronic hearth served to some extent to identify the TV watcher as a patriot, a descendent of Paul Revere and Washington and Betsy Ross, and thus participant in national values. And the portable televisions marketed as luggage encouraged the viewer to feel identified as a vacationer or sojourner, a figure motivated and in motion. Even the images of men in heterosexual on-screen fantasies are projected with approbation by advertisers and cartoonists, who exploited it as an image of containment and public safety.

Yet the sedentary posture of the TV viewer posed a problem neither solved nor suppressed by these other associations. The seated or recumbent figure in front of a TV screen elicited anxieties precisely when positioned in the context of the incipient leisure era. To ask why, we must glance back at the *Life* portrait gallery of prominent Americans, and identify their occupational roles as the president of a manufacturing company, as army general, as operatic bass, senator, museum director, tennis star, and so on. Each is an activist, and each is active in leisure—making, doing, acting on the immediate surroundings.

Each of the profiled Americans, that is, belongs in the familiar tradition of American activism ingrained in myth and school textbook history. The *Life* profiles are congruent with the most cursory inventory of legendary American types—the activist frontiersman, the Yankee peddler, the lumberjack Paul Bunyan, and steel driver John Henry, the immigrant dating from Hector St. John de Crevecoeur's "new man," who earned his very American identity through activism as he cleared, fenced, and farmed his New World tract. And, of course, the military hero, the Washington, Grant, Lee, Pershing, Eisenhower, not to mention the Scouting movement with its quasi-militaristic tradition of physical activism and, too, the figure of Benjamin Franklin tirelessly promoting his own industriousness. Add to these the legendary athletes from John L. Sullivan to Babe Ruth and, in the women's sphere, Babe Zaharias and Alice Marble, in addition to the less well-known types who come into prominence at certain points, like the oarsman and the surgeon, culture heroes of the late nineteenth century, as Elizabeth John has shown in her study of the occupations lauded in Thomas Eakins's paintings. Even the versatile gentleman–scholar–statesman has been figured as an activist, a Jeffersonian involved in architecture, invention, statecraft, agriculture. Elsewhere in this discussion we have seen the extent to which an American contemplative like Henry David Thoreau was portrayed in the Cold War era as a hard worker, not a figure in philosophical repose but an activist representing the best tradition of the skilled craftsman assiduous in his duty. Even in the *Life* special issue, Thoreau returns for review, identified as "an industrious writer even at Walden Pond," a man who "may have talked about leisure more than [he] enjoyed it" (120). The real Thoreau, the text argues, is no devotee of that suspect state, leisure, but an activist worker at heart.

Quite apart from television, the historically based ideological endorsement of exemplary activism seemed on a collision course with the very idea of a leisure culture. Recurrently in the discourses of mass-magazine journalism and their advertisements, one finds that leisure is denied even in its affirmation. The notion of playing as hard as one works, for instance, when applied to business executives essentially conveys the message that the man in question is relentlessly active, a full-time worker. Within this cultural configuration, the figure of the TV watcher in an upholstered chair, stretched out on the sofa, or enveloped in a recliner designed to position him or her in maximal comfort is aberrant—and even threatening to the ideology of work. For that figure signals physical inertia and thus in body language presents a posture antithetical to a nation with activist traditions—the very traditions that had brought modern, industrial America into existence and into global preeminence. The image of the TV viewer flouted these fundaments of national growth and development. That figure, poised just at the crest of postwar industrial society, threatened to trivialize it.

This is a related but separate issue from the passivity attributed to television. And it is separate from the intellectuals' high-culture disdain for the popular, a subject that Andrew Ross has examined at length. Anxiety about the addictiveness of television, about its trance-inducing properties, has principally concerned the monopolization of consciousness, including Cold War fears of mind control so often expressed in reference to the dystopias of Orwell or Huxley. But here we encounter a separate set of anxieties focused principally on TV watching as sedentary, as *physically* immobilizing. Just as fears of TV trances have been largely fears about an irresistible narcotic technology, so those focused on the vision of a leisure society in the TV age have been expressed much more in terms of the human body and the ideological significance of that body in pose and posture. If physical activism represented national energies, then what national traits were signified by the body seated or recumbent in front of a screen?

Television, queried in this way, forces into visibility the contradictory and irreconcilable American strains of leisure and of activism. It exposes an argumentive bind. For the activist energies to which the postwar democratic hegemony of the nation were attributed come into direct conflict with the articulated goals of that energy as they were represented in the figure of the seated or lounging TV watcher. The society of material abundance and of leisure, achieved through more than three centuries of tremendous activist effort, threatened to culminate into the sedentary figure of the TV watcher.

Although the issue ostensibly concerns leisure, it really ramifies directly and more importantly into the world of work, the world that *Life* represented as one of manufacturers, army generals, operatic basses, tennis stars, and so forth. The magazine had not failed to address itself to this subject. "There is nothing new about mankind's abhorrence of too much leisure," it said:

> No matter how much we talk about the necessity of leisure in an age in which machines are taking over more and more mental as well as physical work, our relatively recent Puritan past screams in protest. Since the first settlers at Jamestown starved, the need for work was firmly stamped upon the minds of the early colonizers. . . . Work gained an aura of religious respectability. . . . We still have a profound feeling that work is good, and that the enjoyment of leisure for more than a short time is vaguely immoral. (120)

Yet work and leisure in the television era were no longer such clear-cut contraries as the *Life* essayist presumes, but rather analogues. Despite state-

Whitey Ford takes a Barcalounger-Break
in the Players' Lounge at Yankee Stadium
Now you sit down in a Barcalounger-Rocker
Watch TV, rock or settle back. The rest is easy

Whitey Ford's Barcalounger is the flare-arm "Clinton" Barcalounger-Rocker. About $195.00 at fine stores

BARCA\QUNGER®

The best known and most ————— respected name in reclining chairs

Though shown in posture of relaxation, the New York Yankees pitcher
is surrounded by insignia of his workplace activism.

ments like the one above on work, certain social scientists investigating the
conditions of work and of leisure in postwar America had begun to see the two
as remarkably similar, virtually interchangeable. A 1962 study, *Of Time,
Work, and Leisure*, acknowledged that "leisure could not be treated apart from
work," and it described a modern United States in which "work for the
majority is sedentary and done indoors":

> It is clear that American free-time activities are for the most part done sitting,
> in company, indoors, and at home. Work for the majority is also sedentary and
> done indoors. In these categories, then, the relation of free time to work or vice
> versa, if there is one, is not opposite but similar, along the same line, except
> that work, in contrast to free time, is done away from home. (Heckscher,
> "Foreword"; De Grazia, 183)

This observation, rare in candor its own historical moment, is crucial to an understanding of why it is that television was so problematic within the idea of the incipient leisure era. The figure of the TV watcher, as the statement implies, was really an approximation of the figure of the modern American in the workplace—no longer an activist but a sedentary, immobilized individual. As this analyst remarked, "Machines have already done away with much of the need for muscle power in work . . . . It's a rare bird today whose job flexes his wings. Men in the United States do very little lifting and moving." This analyst assumed that, based on current trends, "work will become less muscular and more sedentary than before," and he predicted atrophied muscles "except perhaps those involved in eye and finger movement." United States working men "start and stop things . . . learning to watch processes and being ready to press buttons" (De Grazia, 297).

The sedentary TV watcher became, then, a version of the American working man, not only the blue-collar workman but the middle and upper-middle class management "man in the gray flannel suit" (a figure made famous by Sloan Wilson's novel of that title), and the "organization man," as William Whyte called him in his sociological study. The composite drawing of all these working men was the seated figure. Transferred into the habitat, he becomes the recumbent TV watcher, and he is not only relaxing or resting from an exhausting day of labor but continuing into leisure hours the sedentary pattern of his work life—and of the commute in his car. As of the mid-1950s, commentators observed that the United States had more TV sets than bathtubs, and that a working person spent seven-eighths as much time before the TV screen as spent at work (Wilson, 290). The explicit juncture of the workplace and the TV place in this comparison was not coincidental. The seated figure before the screen had become an unacknowledged but powerful, even haunting tableau of the modern American person at work.

As "worker," moreover, the TV viewer stood in stark contrast to his activist American predecessor, exposing himself and his modern work place as diminished and degraded. The implied iconography not only of the impending leisure era but of the contemporary work place was thus the American man sitting in front of a television receiver. The advertising discourse for the TV lounge chair featured men in work clothes and thus promoted the idea of relaxation earned from a day at work. Yet analysts committed to an ethos of activist work and recreation regarded TV watching as a sign of inertia. As one said, "A television show may be absorbing for the viewer or dull; in terms of physical movement there is little activity to sitting in a darkened room, even for the eyeballs, except when our viewer gets up to get a can of beer from the refrigerator" (De Grazia, 182). This posture of physical immobilization is the one the *Life* editors and writers felt the need to suppress, because to them it came too close to replicating the conditions of the middle-class work place.

It is ironic, considering the later twentieth-century electronics revolution, that the seated figure before a screen should have been so ominous. The irony doubles on reflection that the computer age, which put much of the U.S. work force from secretaries to architects to engineers to reservations clerks and security guards at work stations behind VDT screens, occurred just at the time when international competition banished visions of utopian leisure altogether. By the 1980s–1990s, as James Schwoch remarks, the American male professional worker was not only presumed to inhabit an electronic work place, but his leisure hours were represented as intellectually committed to that space as well—at least according to the discourse of the telecommunications industry, which portrayed the solitary male exercising—rowing, swimming—all the while preoccupied with work (8–9). Such images of physically strenuous leisure are really intended to prove work-place activism in the electronic environment. The construction of social reality prevalent in the United States from the 1950s–1970s, however, admitted no images of workers seated behind screens, which, at that time, were synonymous with leisure hours, entertainment television.

And the leisure–television–work place conundrum of the 1950s–1970s was especially masculine. Women, most discourses on domesticity asserted, had traditional homemaking duties to attend to throughout the daytime hours. In the closing years of World War II, as Maureen Honey has shown, agencies of governmental propaganda, in their enmeshment with the editors of large-circulation slick and pulp women's magazines, encouraged their readers to commit themselves to postwar lives as homemakers. In the forthcoming peacetime, as magazine fiction and advertising repeatedly told her, the wartime woman worker was to be "the embodiment of peace: a tender, nurturant family woman." Thus "the domestication of the war-working heroine was most marked in stories in which [she was] . . . shown yearning for domesticity, . . . conveying the message that war workers were homemakers at heart" (98–99). Advertising texts sometimes suggested that television could speed up the work process, rewarding the efficient woman worker with a relaxing program. But other Cold War-era texts presuming the woman to be at home all day do convey a suggestive anxiety that the homemaker, drawn to television during workday hours, might neglect her household chores and become a sloven, and cartoon humor especially exploited this point.

Two writters, Betty Friedan and the novelist Evan S. Connell, Jr., challenged the Cold War ideology of the homemaker. Their texts emphasize the underlying vacuity of the homemaker whose life was so typically camouflaged in hyperactivity that anxiety about leisure seemed unlikely to arise. Connell's novel, *Mrs. Bridge* (1958), set in the pre-TV era of the 1930s but written in the

1950s, is a devastating view of the American wife–homemaker who struggles to fill up her days. Connell's Mrs. Bridge is a portrait in affluent suburban WASP emptiness denied and evaded in a cloak of activities. Friedan's complementary, now-classic review of the mystification of the homemaker in the decades she calls "the era of 'Occupation: housewife,'" emphasizes the extent to which the women's culture industry, especially the women's magazines, sought to portray housework as a demanding full-time occupation. Though the suburban housewife of the 1950s–1960s was "afraid to ask even of herself the silent question, 'Is this all?,'" Friedan shows that busy-ness itself was a corollary of housekeeping and foreclosed considerations of leisure. "They were so *busy*," she writes, "busy shopping, chauffeuring, using their dishwashers and dryers and electric mixers, busy gardening, waxing, polishing, helping with the children's homework, collecting for mental health, and doing thousands of little chores" (52, 237). Friedan applies C. Northcote Parkinson's law to postwar decades, "reformulated for the American housewife: Housewifery Expands to Fill the Time Available, or Motherhood Expands to Fill the Time Available" (240).

Advertising images targeted at homemakers in the incipient leisure era began with the premise of their busy-ness and attempted to persuade the American woman, presumed to be middle class, that television was unrelated to her janitorial tasks, that instead it was as elegant as an evening at the theatre—that in fact it *was* such an occasion. Numerous advertisements worked to make that connection. They showed couples in evening attire gathered in their living rooms as if in a private box at the theatre, and gazing in rapt attention at on-screen ballet, opera, or drama from the legitimate stage. Television in the living room was thus offered to the housewife as an excursion out of the household and into an expensive private box for an experience of high culture. With television, the housewife was not an isolated suburban drudge but a threatre-goer, participant in urban social forms of high-culture night life (see Spigel).

On the masculine side, the marketing task was more complex, given that images of work and leisure needed to be kept separate, even when the conditions of both were merging into one, and given also that the man needed both to be shown sitting comfortably and yet must not be allowed to see himself mirrored in magazine illustrations as a sedentary figure. The advertisers and manufacturers of TV receivers, accordingly, constructed their television viewer in terms that denied any work–leisure analogy. Their terms really affirmed television and its viewer as integral with national activism. Faced with the promotion of a product requiring sustained attention by individuals in a more or less sedentary posture, the advertisers needed nonetheless to appeal to

Dumont ad, 1949.

those individuals on a basis that denied certain connotations of that posture. They needed to endorse the sedentary but simultaneously to deny it.

Thus, in the masculine realm, advertisements of television employed deflective strategies. Like the *Life* editors, the marketing workers suppressed all the disquieting associations linking television with the work–leisure analogy. They went further, to deny that TV watching was a sedentary pastime. In general terms, they promoted TV sales and use in three ways, first and most obvious, by showing on-screen images of action and mental intensity, typically moments in sports in which a football is passed as defenders rush in, or a baseball bat is flung aside as the hitter makes contact and starts racing toward first. The male viewer was thus encouraged to identify with the on-screen action of the playing field. As a sports fan, the male TV watcher is an activist.

More subtle are the two other strategies, beginning with those repositioning the viewer and television in new contexts. These move the TV receiver outside the environment of the habitat that is so filled with upholstery, the symbolism of the sedentary. Portable receivers, of course, had made outdoor TV environments plausible, and advertisers were quick to show receivers brought onto porch and patio in good weather. But marketing worked beyond that level of domesticity, to persuade the viewer that he and his TV set occupied a far

Television portables in baseball stadium seats.

wider—and activist—realm, that he was only ostensibly in his customary habitat. For instance, the Panasonic Company, presenting its array of portables on baseball grandstand seats, in 1976, invited the American man to see himself as a figure in the stadium. Such ads, interestingly, endorse the idea of spectatorship, thus giving the viewer permission to be a viewer—but of a certain kind, the man who "fights his way to a seat in the stands . . . and then stamps . . . and shouts himself hoarse" (De Grazia, 182). This figure is the spectator as activist in the outdoors and a participant in decades-long American athletic traditions. These texts tell us that to be a spectator is very different from being sedentary, that to sit is by no means to be consigned to the sedentary.

General Electric's 1960 ad for its "Blue Daylight" TV receiver urges this identification with outdoors activism. At first, the receiver appears to be a picture window onto the offshore scene of sailing. But the onlooker–viewer is not confined to any windowed interior space. The ad says that you are there, at sea, sailing. It makes explicit what most other ads only imply, that the viewer is the voyager, a recreational activist. At first glance, these images seem to say that portable or newer lightweight console TV models can be taken into the wider world with convenience and ease. The length of an electric cord is no longer a tether to the habitat. But the underlying and, given the anxiety of the leisure age, the more important message is actually quite different. This message says that the viewer who is watching television in the habitat is not sequestered there, is not a sedentary figure at all, but is actually at large in the world outside, at the sports stadium or the seashore, for that matter anywhere in the world. The viewer and his TV set are recontextualized à deux as the activist American.

The most cunning advertising discourses in this regard were sponsored by the Television Information Office, an industry trade organization promoting television across the networks. In the mid-1960s, the years when jet passenger travel, begun in 1957, was nearly a decade old, when John F. Kennedy's presidency had set new standards of physical activism, with the president himself regularly shown sailing and playing touch football with family and friends, and when the U.S. space program was sending men aloft in rockets— in these years the TV viewer was invited to enter "The Time Machine" for televisual transport across the entire world, including the "Mediterranean of Ulysses" and "the big wide world of Kenya." Like a pilot or astronaut, the viewer was constructed as a global traveler urged to "take the controls." "If you . . . operate them selectively," the text promised, "television will take you anywhere you want to go." Like the pilot or astronaut, this traveler was a seated figure—seated but never sedentary. As viewer, he would be manning the controls, in charge, guiding himself in the forward thrust of a voyage. He

General Electric ad, 1960.

was the space-age explorer, the commander in the cockpit. The Zenith Corporation promoted its remote-control tuner in just these terms as "Space Command" tuning (*The New Yorker*, 41 [50] [Jan. 29, 1966]: 79; 36 [9] [April 16, 1960]: 79).

Even if this were to be an explicit armchair voyage, the leisure-hours pilot would be positioned in his comfortable chair in careful pose, like the man in the 1966 advertisement that said, "If you want to see the world this February, stay home." Again, television will "take" you internationally to Greece, West Berlin, Moscow, the Netherlands. But the active verbs in the ad text make the

Zenith ad for "Space Command" remote tuning, 1960.

Puts Dad up front...

LA-Z-BOY® for Father's Day

The ad image says this man is anything but La-Z.

TV viewer an activist: "You'll *explore* . . . *travel* back in time . . . *take in* a concert . . . *circle* the globe . . . *tour* the world" [emphasis added]. The man shown doing so, though seated and physically relaxed, is the very vigorous American man, far from sedentary. And so is the working-class man shown in the ad for the La-Z-Boy Chair, the very name of the product counteracted in the image of a working man sitting forward in his La-Z-Boy recliner as if he were "up front," as the text says, at a sporting event. His beer and sandwich forgotten, he prepares to spring to his feet to protest an umpire's or referee's call. No lazy boy he, for every line of his posture, from open-mouthed protest to hand gesture, announces physical and mental activism. And the idea of masculine control continued into the 1980s, in which the Sony Corporation, advertising its Betamax VCR, invited the male consumer to "experience the freedom of total control," just as Turner Broadcasting showed the male viewer of its Cable News Network as a Napoleonic figure who "got control of the world." Control, mastery, world conquest—these are the corporate messages at work to counteract notions of the immobilized male viewer (*New Yorker*, 41 [Jan. 29, 1966]: 79; *Life*, 72 [June 2, 1972]: 69; *Time* [Nov. 3, 1980]: 76–77; [Nov. 13, 1984]: 37).

The lightweight with a lot of guts.

With all the contenders in its size and weight class, we felt we had to equip our new 12-inch* color portable with something special. Something that would let it stand up to any competitor and walk away a winner.

So we took the guts that make our big sets so beautifully automatic, and we worked them into our little Brighton. Pound for pound, inch for inch, we made it the most automatic color portable going.

Space was a problem. So we got rid of a lot of hot tubes and replaced them with a cool 46 Solid-State devices. Which made the Brighton lighter on your feet. And also made sure the repairman wouldn't become a new member of your family.

After that, a lot of things happened automatically.

Like automatic degaussing. That electronically keeps phony color from creeping into a black and white picture.

Like an automatic color-broadcast indicator. So you won't turn blue trying to coax color out of a show in black and white.

There's a special control circuit that automatically gives you true color tones. City or country. Room to room. Day or night.

And another piece of automatic magic that keeps cars, planes, even your neighbor's power tools from messing up your picture.

And, on the premise that you'd prefer to spend more time watching than monkeying, we made it so you could pre-tune each channel for the picture and sound you want. Then it's automatic.

And you don't have to wait for the set to warm up. Something called Speed-O-Vision gives you quick picture and sound.

So if you're looking for a real big-set performance in a size you can waltz around, step into any store we permit to carry the Panasonic line and ask to see the Brighton. Model CT-21P. A lot of savvy went into making it as good as it is.

And a lot of guts.

*12" measured diagonally. Picture simulated.

PANASONIC.
200 Park Avenue, New York 10017
For your nearest Panasonic dealer, call (800) 243-0355. In Conn., 853-5693. We pay for the call.

Panasonic ad, 1968.

Panasonic ad, 1977.

Just as this cluster of images displaces the very idea of the sedentary in favor of activism and space-age exploration, so too do yet another set of advertising texts. They also confront and refute the idea of the sedentary, but they do so within the habitat. Typically, they show sports equipment, for instance, a football bursting through the TV screen into the arms of the surprised but pleased viewer. Amid the upholstered furniture these texts, too, argue that the American man at leisure is actively participant in the onscreen world which is correlatively activist.

Essentially, such images break the frame of the TV screen, disrupting the meanings of the soft rounded corners that under other circumstances denote safety and security. Repudiating the idea of the sedentary, they convey the message that the viewer is a latent activist, a man of potential but untapped action. Suddenly, to the viewer's amazement, the on-screen kickoff sends the ball into his den. His relaxed arms can now spring reflexively to action. In the next moment, he will become a player. And that is the point of such advertising texts, that at heart the seated TV watcher is really a teammate–activist only temporarily on the sidelines but ready at any moment to join the action. He may appear to be sedentary but underneath is kinetic in mind and body. He represents, in short, all the occupational activism by which the nation has defined itself over centuries. He is not at heart a man of leisure, just as the United States is not at heart a nation of leisure, social theorists' prognostications to the contrary. The ad texts say that the activism by which the nation has defined itself still exists in essence, represented in the figure in the armchair. At any moment he will be summoned to action and will meet the challenge, proving that to be sidelined is not to be sedentary. The figure of the armchair activist functions as an anodyne against anxieties about American work and leisure.

It reveals a certain paradox too. Even as futurists hailed the advent of the age of leisure, itself a testimony to the triumph of corporate capitalism, the very discourses ramifying from the corporations vigorously denied leisure validity in the United States. The corporation ads for TV receivers, the trade association messages, the reports on work and leisure authored by representatives of cultural and social institutions all reach consensus on one point, namely, that leisure is antithetical to the ideology of the United States. All their discourses resist it; all make strenuous efforts to deny its actuality. Their denials ring most tellingly in these texts interplaying leisure, labor, and the La-Z-Boy.

Drugs, Backtalk, and Teleconsciousness

*"There are people who don't like TV. And they have reasons.
. . . It has a droning, mindless, addictive quality. It's passive. It
doesn't stimulate.*

Advent Corporation advertisement, 1977

*After her niece leaves, Opal . . . nestles among her books and
punches her remote-control paddle. . . . On TV, a 1950s con-
vertible is out of gas. This is one of her [music video] favorites. It has
an adorable couple in it. . . . They look the way children looked
before the hippie element took over. . . . The boy is dancing
energetically with a bunch of ghouls who have escaped from their
coffins. Then Vincent Price starts talking in the background. The
girl is very frightened. The ghouls are so old and ugly. That's how the
kids see us, Opal thinks. . . . This is a story with meaning. It
suggests all the feelings of terror and horror that must be hidden
inside young people. And inside, deep down, there really are
monsters.*

Bobbie Ann Mason, *Love Life*, 1989

These two texts present irreconcilable versions of the TV viewer. The first is a
stupefied addict, the second a figure stimulated by and actively engaged with
the on-screen image. The first is an abstract type, to be sure, and the second an
individual fictional character, a retired school teacher watching the video,
"Thriller," in which the Michael Jackson character metamorphoses into a
wolf man. But what strikes one about the two juxtaposed texts is that each
repudiates the other, really forbids the other. The two representations of the

TV viewer are in conflict—more than that, irreconcilable. The first constitutes that individual as dependent, addicted, intellectually vacuous as he or she succumbs to the mindlessness of television. Television itself is defined as mind-numbing, addictive, utterly passive, and its viewer is figured as the incarnation of these very traits.

Not so the second representation of the viewer. That figure finds much onscreen worthy to engage the mind. Being a retired teacher shown watching television while sitting among her books, this viewer claims membership in the educational establishment, and we see her in the act of interpretation. The MTV video is "a story with meaning," affording her insight on youth's perception of age and on intrapsychic terror and its repression. Far from passive, this viewer expends interpretive energies that are richly rewarded. Television proffers a text pregnant with meaning, and this viewer meets the challenge in a state of intellectual activism.

When one juxtaposes the two paradigms of the TV viewer, however, it is immediately clear that they inhabit separate, opposed worlds of representation. The teacher is figured as mentally alert, and the video is sufficiently stimulating, sufficiently yielding to her intellectual inquiry that the very notion of the viewer as stupefied addict would be bafflingly anomalous. Yet given the terms of the Advent Corporation advertisement, the teacher's very act of interpretation is impossible: the corporation's discourse argues that television is so mentally numbing that no viewer would be capable of such interpretation even if television were to make it possible, which it does not, being "droning, mindless, addictive."

These irreconcilable representations compel attention to the ways in which the viewer—or, to be precise, the viewer's mind—has been represented in the TV era. The TV viewer, as we have seen, is figured as a hearthside family member and a sojourner, and is also a figure whose body language has been culturally symbolic in a context of work and leisure in the United States. In all these formulations, the mentality of the viewer is implied by the social role prescribed by the discourse. The public consumer of the hearthside image is asked to suppose a mentality of familial harmony, while that same consumer is invited to presume the portable TV viewer to be cheerful, playful, on holiday. At the same time, the viewer settled back in the TV recliner chair is shown in ad texts to be mentally active and alert. All of these visual images, then, imply certain states of mind. The very physical posture of these viewers is intended to suggest a mental posture as well.

Here, however, we focus especially on those discourses that concern the viewer-as-mind. Representations of the TV viewer's consciousness, from stupefied addict to *engagé* intellect, have persisted since the 1950s, disclosing both social anxieties and faith in intellectual autonomy. In recent years, additionally, we have begun to see representations of a new kind of TV-era mentality, a teleconsciousness, that differs markedly from the first two formulations of the viewer as addict or as resister. All of these representations coexist concurrently in public discourses, although for clarity it is best to see them in sequence.

The tableau of the mesmerized viewer has persisted throughout the TV era, advanced by critics whose paradigmatic viewer is constructed as a form of still life. Critics have warned against a nation of robotized individuals and their families living hand to mouth out of cellophane bags in front of the television, itself an anesthesia chamber of corporate capitalism. The critics fear the addiction of men and women so narcotized by television that they become immobilized, apathetic, robotic. In 1948, *Life* magazine warned that television might soon be "hard on the housework" and pronounced that situation a "horror," and in the 1960s McLuhan warned, "We have leased . . . our eyes and ears and nerves to private interests" ("Salute to Television"; McLuhan, 73).

The industry, attempting to overcome such fears, has acknowledged their prevalence, as in the Advent Corporation advertisement quoted above (*The New Yorker*, [Dec. 12, 1977]: 181). But for the most part the industry has scoffed at such images of actual ennervation, its public discourse portraying its viewer as a sovereign being whose satisfaction it pledges itself to ensure, even as the critics rejoin that the viewer is also a market and a consumer manipulated into unslakable thirst for goods and services and ever more television. The discussion continues, the viewer formulated by commentators outside the industry as a figure of entranced passivity.

The slang terms for the TV set and the viewer—boob tube, idiot box, couch potato—reinforce this view, indicting both the audience and the apparatus. "TV hermit," "lumpen lug," and "addict" ring in denunciation (Gerbner and Gross, 42; Markfield, 28; Reynolds, 24). The slang reveals the anxiety of self-dislike, made graphic in two *Saturday Evening Post* cartoons of the 1960s, the first of which shows a man sitting motionless in a chair facing a wall while his wife directs the TV repairman to put the set "over there in front of my husband." We understand that while the set has been out for repair, probably for days, the stupified husband–viewer has sat in his TV chair staring at a blank wall, catatonic. Television has done this to him, just as it has turned the man in a second cartoon into a turnip. (*The Saturday Evening Post*, 234 [Feb. 25, 1961]: 84; 238 [April 10, 1965]: 80).

In part, the vehemence of these images is attributable to social anxieties about narcotics. A society focused since the 1960s on drugs would predictably extend its angst about drug addiction to television, and we duly note the titles, *The Plug-In Drug*, "So You Can't Kick the TV Habit?," "The TV Addict"

(Winn, Reynolds, Tennesen). Imagery of narcotics threads its way through much public discourse on television. "Some people . . . ignore TV because they are afraid of getting hooked," wrote a *Time* magazine journalist in 1968. "What is to be done with parents who are themselves television addicts?" asked a writer in 1964. Television could "become the worst cultural opiate in history, buy and corrupt all talent, and completely degrade the sensibility of the country," one critic wrote in the early 1950s. And recently the journalist Russell Baker, looking back at the early 1950s, remarks, "There were signs that television, like whiskey, was soon to become a consoling vice of the poor." And newspapers continue to feature such articles as "How Viewers Grow Addicted to watching TV" (*Time*, Nov. 8, 1968, p. 98; *Saturday Review*, Dec. 5, 1964, p. 42; Willingham, 117; Baker, 139; *The New York Times*, Oct. 16, 1990). One television producer commented, "A TV addict who [fiercely criticizes] the industry is like a narcotics user who blames the pusher." He goes on in direct address:

> Face it, Angry-Viewer-Who-Is-Hooked: If you have become a wasted blob from TV dinners, if your children have become bleary-eyed zombies, it is *you* who must exercise self-control, it is *you* who must be the police-man. (Aurthur, 184–85)

Here, of course, the addict is exhorted not only to reform but to become the uniformed disciplinarian and agent of the state.

But this anxiety about drugs seems to be only the expression of a deeper long-time fear, namely, that television, unlike its predecessor, radio, commands the totality of individual consciousness. Recently, in 1988, Mark Miller, invoking an Orwellian world, argued that television is a totalizing commercial environment cocooning the viewer "against all contrary evidence out in the streets and beyond" (12). But the fear of the monopolized consciousness has been a constant throughout the TV era. "The image . . . demands complete attention," warned Gilbert Seldes in 1937. "You cannot walk away from it, you cannot turn your back on it, and you cannot do anything else except listen while you are looking" (535). "Television is unquestionably a much more arresting medium than radio," wrote another critic, in 1948, presenting his fears as the unassailable truth that would continue to undergird TV criticism for decades: "It's a medium that monopolizes entire attention when it is on, and prevents one from doing anything else constructive" (Utley, 138). "We suspend virtually all the accumulated habits, chores, conversations, recreations, social amenities and thought patterns of pre-TV peoples," said yet another voice, "and huddle together in the dark, absorbed in the fleeting dance of electronic impulses across the face of a cathode ray tube" (Faught, 413). This premise makes consciousness itself a univocal entity and television the

Accompanying a 1977 *Saturday Evening Post* magazine essay on television as narcotic, this drawing shows hypodermic needles as electric plugs. Reprinted courtesy of © Curtis Publishing Co.

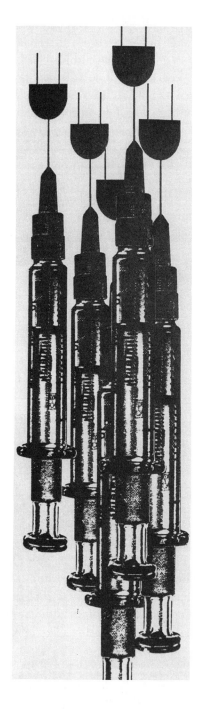

nemesis of constructive action of every sort, "constructive" presumably used in a positivist sense. The viewer is by definition immobilized, and thus we hear the anxiety expressed in terms of addiction, trances, soporifics, vegetative states. Urie Bronfenbrenner uses the image of sorcery: "Like the sorceror of old, the television set casts its magic spell, freezing speech and action, turning the living into silent statues so long as the enchantment lasts" (in Winn, 40).

Recognizable in all these images is the Cold War anxiety about the mind-control of a centralized totalitarian state and about the quiescent conformity popularity attributed to Communism. In the postwar decades, the perceived external threat of nuclear attack by the Communist Bloc made the idea of the American home especially compelling as a haven (May, 3–15). Because radiation, however, could penetrate the solid walls of the home, fears arose about the very idea of invisible forces infiltrating the most secure centers of personal and social life. Television cathode rays and ionizing rays from nuclear fission and fusion were not conjoined in public discourse, but the very idea of invisible, penetrant radiation from an enemy heightened paranoia about invisible invasion and conquest. Senator Joseph McCarthy and the House UnAmerican Activities Committee operated on the premise of the unperceived infiltration of enemies, all the while the perceived threat of invasion by the Soviet Union made it unpatriotic to oppose U.S. nuclear testing, whose radioactive fallout nonetheless became a source of public anxiety—as if the radiation sickness caused by American nuclear testing was but a tiny sample of the real horror the Communists would inflict in the next war. The real nuclear threat was thought to be Communist, though ironically the nature of that threat was made clear from reports of radiation sickness from American weapons tests in the Pacific and in Nevada. David Bradley's best-selling *No Place to Hide* (1948) argued that there is no defense against radiation, and though government propaganda worked to downplay the danger in the mind of the lay public, by the mid-1950s the press was steadily reporting the perils of radiation (see Caufield, 89–132).

The very idea of dangerous or subversive infiltration evidently extended to television. One set of discourses attempted to insulate consumers from fears of outside threat by emphasizing that television was the modern family hearth. Another set, however, expressed the anxiety that television itself was the force infiltrating to the very center of life, destroying human will and, therefore, imperiling the cherished American ideal of individual freedom. The Beat generation novelist, Jack Kerouac, in *The Dharma Bums* (1958), showed how television was thought to function within this Cold War outlook when he described each suburban American living room "with everybody looking at the same thing and thinking the same thing at the same time . . . house after house on both sides of the street each with the lamplight of the living room . . . and inside the little blue square of the television, each family riveting its attention on probably one show" (32–3, 83). Even in 1990, one critic compared television to a subversive attack by hallucinagens: "one of TV's best tricks is to never let on just how much it affects our psychic mechanisms. Like LSD in the water supply, it alters our very standards of judgment" (Davis, 89). The TV viewer, so configured, is entirely lost to the world and to the self as long as the TV set is turned on. With consciousness so monopolized, no one has the will to turn the TV off, so every viewer is epistemically and on-tologically imperiled, held in a state of suspended animation, if not ruined. Narcotics and totalitarianism thus become the Cold War, nuclear-age threat of television as it monopolizes consciousness.

Yet diverse TV-era texts have consistently refuted the idea that viewing is synonymous with passivity. These texts report that TV watching is neither a sign of assent nor of acquiescence. They show, on the contrary, a continuing irreverence for and resistance to television, the signs of which probably go back to the black-and-white days when viewers sometimes turned down the sound in even the most serious and dignified scenes and saw the inevitable result—comic pantomine. And we now begin to have the memoirs that reveal, in anecdotal form, an irreverent activism vis-à-vis the viewer and television. One woman describes her husband as he watched TV news of street demonstrations in the 1960s: "He'll be watching the news, and he shouts at the demonstrators" (Lora, 314). And Annie Gottlieb's interview with a Boston banker prompted his memory of a TV moment in the early 1970s: "The president of the United States got on TV, and guys were dropping their pants and sticking their asses up! They were mooning the president" (44). Politically, this TV backtalk cuts two ways. In her memoir, the former presidential speech writer, Peggy Noonan, recalls that in childhood at a friend's house:

> "I'd hear her mother yelling at the TV upstairs. 'You tell 'em, Bill! Tell that big jerk!" She was watching William F. Buckley, Jr.
> "Facile jackass!" She was warming up on John Kenneth Galbraith.
> "Poseur!" That must have been Jerry Rubin. "Boob!" (14)

These have been the marginal, desultory TV texts, overshadowed by the social projection of fear and anxiety about narcotics and totalitarianism. And where else have they appeared?—often in cartoon humor. Virtually eluding commentary, cartoons have recorded the ways in which the TV viewers engage in resistant backtalk. Line-drawing cartoon humor from the 1950s onward has shown much more directly individuals' interactive, often combat-ive relation to television. Typically, the viewer in these cartoons is cast in an argumentative pose, rebuking or rebutting the on-screen figure whether an

American Airlines ad, 1956.

New Yorker Cartoon, 1970. Reprinted courtesy of *The New Yorker*.

advertiser, actor, or politician. The irate husband, for instance, watching a TV ad, turns to his wife and says, "Listen, just because this Jim Dooley says 'Come on down' doesn't mean we have to go on down." Or a wife says to her husband, "Maybe it's me, but tonight it seems *everybody* on television is lying" (*New Yorker* 42 [March 5, 1966]: 34; 51 [June 2, 1975]: 39).

One notices that, cartoonists' individual styles aside, both the viewers and the on-screen figures in these drawings bear a strong resemblance to one another. In physiognamy, in dress, and in age the two figures mirror one another. No contention in race, gender, ethnicity, class, or generational difference gets inscribed in the drawing. The virtually interchangeable figures assure us that their differences in viewpoint will not be socially disruptive. It is safe for the viewer to repudiate the product or the political message because, the drawings tell us, that viewer shares the same underlying values as the on-screen figure. Their very similarity can free the cartoonist to show anger on the part of the viewer.

Television-related humor, nevertheless, for decades has represented the viewer as an active figure resistant to the on-screen message. These drawings suggest how insufficient is the image of the viewer as silently quiescent before the TV screen or narcotized by it. The cartoons argue, on the contrary, that a critical, judgmental viewer is actively positioned vis-à-vis the self and the on-screen TV world. The humor of such cartoons turns on the very idea of backtalk. The viewer will *not* try the on-screen product, will *not* vote as urged to do, will *not* accept the referee's call. The viewer is an oppositional figure shown in a posture of active resistance.

And as the mid-1960s, the backtalk took a new, technological form as viewers became empowered in ways evidently unforeseen by the manufacturers to TV equipment. The low-cost hand-held remote became the viewer's means of unprecedented control. Not that the industry necessarily envisioned its current use. Ad copy suggests that convenience of picture adjustment and tuning was uppermost in industry's mind in years when, simultaneous to the remote, color sets were being aggressively marketed. "Now, remote control TV that's tuned in before you turn it on," proclaimed a 1972 Emerson Corporation ad emphasizing that a "single touch" of the "permacolor button" eliminates the need to adjust the set. "Controls volume," said one 1965 Admiral Corp. ad—"changes channels, controls color intensity and tint, turns set on and off—all from your easy chair." The emphasis was easy-chair, BarcaLounger convenience. Seated, the viewer could adjust tuning and volume and change channel—of course on the hour and half-hour when the program changed, for *that* clockwork assumption ran so deep it did not need statement. The remote would simply bring the control panel to hand, eliminating the need to get up and down, especially in years when color was so new and on-screen faces tended suddenly to go green or orange (*Life*, 48 [4] [Feb. 1, 1960]: 38–39; 73 [17] [Oct. 27, 1972]: 31).

But, unforeseen by the makers and the marketers, the remote brought the viewer into an activist relation to the on-screen worlds and increased the powers of resistance. With the remote, the opposing thumb became an agent of irreverence, breaking tradition (and inertia) by enabling channel changing fast and often. With the remote, a reverence for, and subordination to, the integrity of the sponsored program itself disappeared. Even viewers shown to be at a low-energy moment are TV activists with the remote, as John Updike suggests in a fictional vignette in which his protagonist, "Rabbit" Angstrom, sits watching television with his teenage son, both of them depressed because the wife-and-mother has left them. "They channel-hop," writes Updike, "trying to find something to hold them, but there is nothing, it all slides past until, after nine, on Carol Burnett, she and Gomer Pyle do an actually pretty funny skit about the Lone Ranger" (*Rabbit Redux*, 29). Even at their most ennervated, father and son "channel-hop" until an on-screen scene merits their attention. This active, channel-hopping viewer appears in stark contrast to the zombielike viewer—the turnip—described earlier, suggesting a multiplicity of discourses, many of which configure the viewer's mind differently.

As for the channel-hopper, that viewer enacts a cognitive process identified in the mid-1970s by Raymond Williams. In *Television: Technology and Cultural Form*, Williams cautioned that TV reviewers were misguidedly, anachronistically operating like drama or film critics or book reviewers, approaching individual programs as "a discrete event or a succession of discrete events" (88). Williams, the British Marxist social analyst with particular interests in the cultural institutions of print, had been a BBC television reviewer between 1968 and 1972, and he became convinced that forms of broadcasting in the TV age were altering perceptual processes. Prior to broadcasting, Williams observed, "the essential items were discrete . . . people took a book or a pamphlet or a newspaper, went out to a play or a concert or a meeting or a match, with a single predominant expectation or attitude" (88). The fundamental expectation was of a discrete program or entity.

But increasingly, Williams found, in the era of TV broadcasting the discrete program has yielded to a structure far more fluid. "There has been a significant shift from the concept of sequence as *programming* to the concept of sequence as *flow*." He goes on: "There is a quality of flow which our received vocabulary of discrete response and description cannot easily acknowledge" (93).

Raymond Williams's identification of "flow" proved a benchmark in differentiating the experience of broadcast television from other forms. Conceding that vestigial elements of discrete programs remain intact in the timed units of a "show," he argued nonetheless that the intervals between these units have disappeared. In American broadcast television the advertisements are incorporated into the whole:

> What is being offered is not, in older terms, a programme of discrete units with particular insertions, but a planned flow, in which the true series is not the published sequence of programme items but this sequence transformed by the inclusion of another kind of sequence, so that these sequences together compose the real flow, the real "broadcasting." (90)

Then, Williams observed that the additional, more recent insertion of trailers and previews of forthcoming broadcasts further contributes to and complicates the flow, and he calls all this "a new kind of communication phenomenon" demanding recognition (91).

The demarcation of flow by Williams has been tremendously influential among scholars and analysts of video forms. Every subsequent analyst of the medium of broadcast television has addressed the concept, some with modification (see Robert C. Allen, E. Ann Kaplan, John Ellis). All uphold the essential idea of fluidity as a dominant trait of television.

The channel-hopper is the viewer of television's flow, not only subject to it but exploiting it aggressively. Under the most ordinary circumstances of TV watching, the contemporary viewer with remote can now discard the undesirable scenes or, contrarily, exploit them, moving among many in montage and juxtapositioning. This viewer becomes a kind of *auteur* creating a personalized program. "Zipping" and "zapping" may refer most often to the viewer's personal editing, to muting or to scrubbing the commercials and tedious segments on VCR cassette tapes—but the terms pertain as well to the "couch potato" as activist sprouting eyes eager for a multiplex of TV images.

In this sense, the couch potato's eyes signal visual appetite, and the remote—a customizing splicer/juxtaposer—feeds it. The media critic David Marc has commented on this in his *Demographic Vistas*, reminding us that the cable converter and remote work together. From the divan, Marc writes, "the viewer is less committed to the inertia of program choice. It is possible to watch half a dozen shows more or less simultaneously, fixing on an image for the duration of its allure, dismissing it as its force disintegrates, and returning to the scan mode. Unscheduled programming emerges as the viewer assumes control of montage" (34–5). Cautioning that the freedom of the viewer-as-consumer has not increased, Umberto Eco also recognizes this "incontrollable plurality of messages that each individual uses to make up his own composition with the remote-control switch" (148).

And the channel-hopping viewer appears in recent fiction, which represents the irreverent interactive TV experience of the eighties. The novelist Trey

This couch potato watches multiple TV images simultaneously in drawing by artist David Bigelow of Flint, Michigan. Reprinted by permission of the artist.

Ellis, in *Platitudes* (1988), structures his novel on several contemporary American rituals, including quiz programs, SATs, restaurant specials (and the first-name waiters manning their cannon-barrel peppermills), and public radio pledging. And channel flipping. Ellis's high-school age protagonist, Earle, reclines in a Stratolounger with dinner on his lap and the TV remote at hand, switching to the "ScanTron-Plus mode" which "begins its automatic ten-second channel sampler":

> *Gosh, she's the most beautiful woman I've ever seen: sexy, sweet, witty, and intelligent too, but what's this? Oh no, she's scratching her head. Must be dandruff. Forget—*
>
> If Lucy didn't realize Ricky hid the gunpowder from the firecrackers he took away from Little Ricky in the pepper shaker, then that means at the ladies' club luncheon, Lucy's *Boeuf Flambe*—
>
> Fire and Ice is like no other synthetic motor oil on the market today, since it is fortified with Polyaminide-80, so will protect your car in an extra-special way far too complicated to explain here on Tee—
>
> "Visionary" is how I'd describe myself, Merv. Know that sounds conceited, what have you, but I feel darn proud to be the first celebrity of note to have his own line of designer Lucite furniture—
>
> "We're movin' on up / to the East Side / to a dee-luxe apartment in

the sky-ayyy. / Aw movin' on up / to the East Side. / We fin'ly got a piece of the pie"—

   Aaaaaay! Tonight at eight o'clock you have an appointment with terror. From the director who brought you such classics as *Bloodbath* and *Eyeballs* comes— (86)

And on they go, making the entire chapter a montage of TV Americana, including local news ("suspects, both black males . . . one of stocky build wearing a gold chain"), MTV, sitcoms and cop show reruns, ads for food, deodorant, and gadgets probably not sold in stores ("this extra-special burgler alarm, smoke, and noxious-fume detector"), baseball, Jazz greats (Art Blakey, Wynton Marsalis), soap opera, Accu-Weather.

At first, Ellis's channel-flipping montage reads like a transcript with just a few incongruities, a motor oil named for a perfume, or "designer lucite furniture." But as the chapter proceeds, his careful juxtapositioning becomes apparent, as do the satirical elements that predominate, the "traveling contraceptive salesman," the figure of "Cranky Robinson, the newest member of the Yankee organization," a product called "Swiney's Au Gratin Potato Food," an ad whose copy threatens, "so if you love your children and don't want some social agency to take them away, you'll spend the extra pennies." The satire and the ad hyperbole blend almost seamlessly, the former just a degree more outrageous than the latter, so that the deliberate literary satire becomes a commentary on the self-satire intrinsic to every genre and format of TV programming. (Perhaps Ellis would agree with the critic, Marc, that the sitcom is the heart of television.) Presuming his readers' familiarity with the multichannel mix, Ellis does what F. Scott Fitzgerald did in the 1920s in *The Great Gatsby* when, assuming his readers' familiarity with the newspaper society column and its lists of socialites' names, Fitzgerald listed a roll call of the guests at Gatsby's parties ("the Leeches," "a man named Bunsen," "Doctor Webster Civet," "the Willie Voltaires," "Clarence Endive," "Newton Orchid" [61–62]). Fitzgerald's names, like Ellis's TV images, shade off just perceptibly from those in the texts they parody, so that reader and viewer are made to see the self-subversive satire intrinsic in the primary forms.

The remote-armed viewer emerges here as satirist and activist, but Ellis's text suggests something else as well. The chapter montage comes around twice to a broadcast of a Yankees game, which indicates that young Earle's remote has gone through a complete cycle of channels and started through them once again. Earle, like the reader, is mindful of the repeated short scenes from the ball game, and, were Ellis to send us through another cycle of ten-second glances, he would doubtless offer additional scenes from the montage, perhaps from *Dynasty* or the *I Love Lucy* rerun. Ellis does not do so, choosing instead to broaden his satire to the edge of the outrageous. But were the novelist to extend the chapter-montage, Earle and the reader would begin to track these programs. We would begin to follow them simultaneously, aware both of Lucy and Ethel and of Blake and Crystal Carrington, not to mention the ball game and other scenes.

By returning to the Yankees game, Ellis implies the presence of a viewer (and reader) whose consciousness can be multiply focused on more than one program at a time. Though *Platitudes* does not develop that point, its implicitness leads us inevitably to another, related issue in viewer consciousness. Just as the narcotized TV addict and the backtalking viewer are inscribed simultaneously (and irreconcilably) in public discourse, so a third formulation on the mind of the TV watcher has more recently emerged. Neither addict nor TV iconoclast, this formulation of the viewer's mentality emphasizes the mobility and pluralism of consciousness.

The discourses on stupefied addicts and the backtalking viewers leave a large excluded middle ground. Where, we must ask, are the texts that might fill in the spectrum of viewer types? Which discourses position the mind of the viewer between the two representational extremes in which that individual is either stupefied by television or resisting it with backtalk? One thinks of the 1986 lyric in "Love for Sale" by the rock group, Talking Heads: "I was born in a house with the television always on," a line that points toward the inevitable representation of life in such habitats. If the passive, trancelike state is to be understood as a projection of sociocultural anxiety, then what is the state of consciousness represented when self and others cohabit with an "always on" television? What is the conception of the interaction? And how is it represented?

We might begin with a 1955 television industry advertisement. "Where did the Morning go?" asks the NBC ad, showing an aproned housewife holding her feather duster, her preschooler at her side. The morning "was a pleasure instead of drudgery . . . the chores are done, the house is tidy, the ironing's finished and put away, but it hasn't *seemed* like a terribly tiring morning." This housewife, according to the ad, watched NBC's programs while ironing. And to read her log of programs watched and chores done is to learn that the TV-watching housewife is no sloven, that the informative programs make her a better housewife and more informed individual, and that her child is being educated as well as entertained by "Ding Dong School."

Yet the ad also acknowledges that two activities, ironing and TV watching, proceed together. True, among human endeavors, ironing may rank with those least demanding of the frontal lobes, and the woman's fixed position at the ironing board with a load of laundry puts her just where NBC wants her, shackled to her tedious task for hours in full frontal view of the TV screen (a version of the viewer as still life). But the image of household chores proceeding *along with* TV watching modifies the images of trances and addiction. It introduces a different representation. In 1950, the TV critic, John Crosby, said, "It's impossible to get anything accomplished in the same room while it's on" (67ff). Yet the ad rejects this position, saying this woman has been doing two things at once, the very state of simultaneous actions belying the notion that television monopolizes "entire attention" and "prevents one from doing anything constructive." This woman has been *both* ironing *and* watching. We have here the acknowledgment of simultaneity. The interactive viewer is one who is engaged in activities both in the habitat and on the TV screen.

And if we suppose that this woman had another child, we might expect her to be the speaker of a humorous, anecdotal magazine essay appearing in 1954 within months of the NBC ad, and addressing the same issue of consciousness simultaneously involved in two, or even three, activities, including television. This mother appears to lodge a complaint that the children rejected television and destroyed their mother's new-found peace once the novelty of the electronic baby sitter wore off:

> The children began to play jacks, read the comics, fight, and flip baseball cards while watching. This still kept them roughly on the site of the television set but hardly immobilized. (Whitbread and Cadden, 82)

These very children, we have been told, had previously sat in silent absorption before the likes of "Howdy Doody" and "Super Circus." Yet now they have taken up other activities—conducted simultaneously. Tongue-in-cheek, the mother pretends chagrin that the widespread theory of viewer immobilization has proven false, since a dose of TV was supposed to anesthetize the rambunctious children. Instead, in some perplexity she sees them play games and read and fight with one another, and intermittently watch TV—open to any of these activities at the same time.

Probably unknown to her, this observation would prove prophetic of a new interactive consciousness in which television becomes one area of attention among others. Consciousness would extend into several activities at once, TV watching being just one of them. These mothers, both the NBC ad housewife and the observer of the children, point toward the kind of interactive consciousness characteristic of the viewer in the habitat in which the television is "always on." The ideology of the narcotized, immobilized viewer, however, so powerfully dominated discourses on the TV watcher that the multiply active viewer went unrecognized for decades. Even so astute a social analyst as Michel de Certeau, locating the many daily life strategies individuals employ to resist the "totalitarianism" of a capitalist–consumer environment, accepts the idea of the TV viewer as entirely passive, "a pure receiver" (31). Only very recently has the concept of mediated communication entered the design of studies on TV watching, as communications experts acknowledge that decades of experiments, so carefully designed to be "scientific," in fact drastically change, erase, or alter the conditions of the actual TV environment. One small 1971 study involving the household TV usage of twenty families found viewers "engaged in a host of activities while the set was on, reading a newspaper or magazine, talking, sleeping, staring out a window," all phenomena "that regularly occur in natural environments yet are suppressed by researchers" (J. Anderson, 204).

Perhaps it is not coincidental that one of the first psychological studies to investigate this multiply active kind of viewer originated, not in the United States but in Australia, and involved children as subjects. The researcher concludes that children around the TV set constitute "the lively audience," for her data verify findings from the twenty American households and the anecdotal observations of the mother whose children fought, flipped baseball cards, and played jacks while watching television (see Palmer). The Australian psychologist finds that while watching TV, young children play with pets, look after brothers and sisters, play board games, make and build things, play with toys, jump and dance, read, do homework, fight and talk (148–51). "The TV babies really can do their homework, watch TV, talk on the phone and listen to the radio all at the same time," argues a guest columnist in a 1990 *New York Times* editorial. "It's as if information from each source finds its way to a different cluster of thoughts" (Pittman). This columnist, significantly, writes under a *Times* rubric called "Voices of the New Generation," positioning himself as one bred to be a "TV Baby."

As of the 1950s and 1960s, then, children would grow up living in this kind of TV habitat, taking it for granted as the natural state of things and cultivating their multicentered consciousness, inevitably to represent it in their own texts. The mother's mid-1950s anecdote, in fact, pointed toward a new state of mind that can be called, to indulge in a neologism, teleconsciousness.

What is it? First, the texts that represent individuals and groups in a setting in which a TV set is turned on reveal a new pluralistic state of consciousness in which engagement in the on-screen realm proceeds simultaneous with engagement in diverse activities in the habitat. Attention shifts to and from the television, including it but not excluding other activities. A 1960 magazine advertisement for Irish Mist liqueur, for instance, suggests something that ads for TV sets would never dare hint—that as soon as you and your friends figure out who the culprit of the TV mystery is, "the television can be disregarded and conversation can ensue" (*New Yorker*, April 9, 1960, p. 95). We notice that the ad does not suggest turning off the set but only turning attention elsewhere in the same environ, all the while the program continues.

Conversely, as other activities proceed, television is continuously available and can be watched intermittently. And by the mid-1970s, Michael J. Arlen, working as the *New Yorker* TV critic, was representing this new consciousness in a deliberately transcriptive format. Here is an excerpt from a sketch he subtitled "A Fable," in which he presents the typical middle-class Thanksgiving Holiday in contemporary America:

> Mother and Aunt Sarah cleared the dishes after the main course. Aunt Sarah asked Eddie Jr. what he was doing underneath the table. Eddie Jr. said that he wasn't underneath the table. Mother brought in the pumpkin pie and the dessert plates. "Billy Van Edwards was out there in the flat, but the pass was wide," the announcer said. "That was great protection, Lew." And "Now Bengston gets back to kick. He hangs it high. Rowen is under it. He's hit hard by Delmonico and Harry Widcombe" . . . Mother called to Father that his favorite dessert was on the table. Father and Uncle George went back into the dining room. Uncle George said that that was what he would call a piece of pie. Father smacked his hands. Mother said that she hoped that Father's team was ahead. Father said that football was not so big a deal with him that he would pass up a piece of pie like this one. . . . Grandmother said that she was reminded of the time she once visited her sister in San Fransisco. Her sister was married to a man who had a telescope on the back lawn. . . . A cheer emanated from the television set. Father said he would be back in half a moment. (90–91)

The entire sketch proceeds in this way, and Arlen's fable, told in a style of the dreariest of elementary school basal readers, is intentionally his primer on the trivialization of a sacred–secular holiday by the intrusive television. The family gathering in the age of television is reduced to conversational non sequiturs and palliatives. (Even the children eventually fall asleep in front of a small black-and-white portable.) No reader could miss the message of Arlen's fable. We are to experience the poignance of the lost, pre-TV world of the Norman Rockwell Thanksgiving, to regret and deplore this newer version, implicitly to resolve to restore the holiday by having next year's Thanksgiving without TV parades and football. Yet Arlen, consciously but unwittingly too, is transcribing the state of teleconsciousness already entrenched in the United States.

In this formulation, thought itself changes, becoming multivalent. The individual is cognitively functioning in two or more places simultaneously, in affairs of the habitat and in those of the on-screen world, assigning primary attention alternately to one or the other. Throughout this study, we notice the many instances of children doing homework in front of television, in John Updike's *Roger's Version*, in Betsy Byars's *The TV Kid*, in Allan Bloom's image of the "thirteen-year-old boy sitting in the living room of his family home doing his math assignment while . . . watching MTV" (74–5). In each example, the adult narrator takes a dim view of this, considering it to be *divided* attention in which the only valid priority, the homework, suffers from the distraction of television, just as in Arlen's sketch the holiday ritual suffers from the same distraction. For the narrators, there is but one legitimate locus of attention against which the television is a distraction or an interruption. Each narrator, then, presents a case of *divided* and *interrupted* attention.

From the school child's point of view, however, it is not attention divided, not an either/or state, but one of multiplicity as the mind turns simultaneously to several centers of attention, constantly prioritizing and reprioritizing among them. The on-screen world may be compelling, even deeply absorbing, but only intermittently so as the mind reassigns itself new centers of attention.

The 1970s–1980s have brought numerous texts representing this tele-consciousness, showing the television *on* as individuals go about their tasks. And it is helpful to examine just two of them at some length to see the actualization of this TV-era teleconsciousness, bearing in mind that the authors of these texts are the now-grown-up children of the generation that once played jacks and flipped baseball cards in front of the TV screen, which is to say that they are writers of the television second generation. They are not in the older camp of the Michael J. Arlens for whom the television "always on" is necessarily interruptive or disruptive, or invasive. Instead, they understand the constant of television to be one element in the natural order of things, and they see it as both realistic and, as writers, also grasp its potential as a literary resource. From Lee Smith's *Black Mountain Breakdown*, a southern novel, and from Frederick Barthelme's *Second Marriage*, a novel of postmodernist sensibility, to take just two cases, we encounter the representation of tele-consciousness, of the ways in which consciousness in the TV era is bicentered or even multicentered, with a continuous reassignment of priorities of attention.

In *Black Mountain Breakdown*, we see the protagonist, Crystal, poised between childhood and the imminent high school years as she has her hair streaked in a kitchen during an episode of the TV soap opera, *Days of Our Lives*. In a straight-backed kitchen chair, "surrounded by women," her mother and her cosmetician aunt, Crystal dons a tight bathing cap and endures an "incredibly painful" rite of passage, a process in which the older women relatives work crochet hooks through the rubber cap, "jabbing them down into each hole and then jerking, [followed by] the slow pull up until a whole long strand of hair emerges."

The makeshift beauty parlor aside, the novelist presents this as a typical, ordinary summer day in a small town kitchen, framing her scene with three women, one with an infant, the other two plying the hooks, all three gossiping as *Days of Our Lives* airs, "and every now and then the women pause in their conversation to see what's going on in Meadville and then resume, jabbing and pulling and talking." Essentially, this clause is the writer-to-reader guideline for the entire scene, in which life in the kitchen will proceed along with the TV program, which is seen daily in this household. The novelist announces the terms on which she will present the scene, those of contemporary verisimilitude. Insofar as the TV program commands consciousness, that will be represented. So we begin, with Crystal biting her lip in endurance of the painful streaking:

> "Crystal, honey, turn your head over this way a little bit," Neva directs. Crystal turns her head. Her eyes are on a level with Neva's armpits and she sees a wide wet patch of perspiration on the blue uniform. Crystal sniffs but it doesn't smell. . . . Neva . . . is a big-boned energetic woman. This year her hair is auburn red.
>
> "I cannot go on like this!" says a beautiful woman in Meadville, clutching at a doctor in a hospital corridor. "Let me tell Gregory! We cannot live this way. Always meeting in secret—the motels, the deceitful lies. It's killing me."
>
> "You must calm yourself, Karen," the doctor says dispassionately, looking quickly up and down the hospital corridor to see if anyone has heard. . . . We will discuss this at another time. Just now there is something I have to tell you. It is not good news. I want you to brace yourself. Are you ready?"
>
> Karen gulps and bats her eyelashes tremulously. It's clear that she is not ready at all.
>
> "The test results from your physical have come back, darling, and I regret to say they indicate that you may have a—"
>
> Organ chords crash and a commercial for Oxydol appears.
>
> "Well, shoot!" Lorene says.
>
> "I bet she's got a malignant tumor," Neva says, jabbing.
>
> "She might," Lorene admits. "You know she hasn't been feeling so good. But I thought it was just nerves."
>
> "Maybe she's pregnant," says Susie, tilting Denny's bottle up so he can get it all.
>
> "I don't think I can stand this anymore," Crystal says suddenly, surprising everyone including herself. Usually she has such nice manners. "It really hurts." (38–39)

The women decide to give Crystal two aspirins, and the narrative moves into her private thoughts, a meditation on the childhood pleasures soon to be left behind: "long days out riding her bike, reading, going to the movies. . . . Just sitting, mostly, in different places: by the river, on the back steps, on Agnes's front porch. Mooning, her mother calls it" (39). Crystal rejects the parental pejorative on her summer interlude, renaming it affirmatively, "biding her time," after which the fictional consciousness moves back into the kitchen, where the young mother puts down her sleeping baby to escape to the grocery store, the escape itself prompting conversation on mothers cooped up with infants, unhelpful fathers, the quality of husbands in general:

> "Edwin could have done a lot worse," Lorene remarks, and Neva says, "I guess so."
>
> "What do you mean, you spent the night at your friend's house? You don't have any friends. That's a miserable lie," cries Mrs. Bennett in Meadville. "Sandra, answer me."
>
> Sandra, a long-haired skinny teenager, rushes up a flight of stairs. Mrs. Bennett goes into the bathroom and takes a pill.
>
> "If you give them an inch they take a mile," Neva says to no one in particular. "OK, honey," she says to Crystal, "now shut your eyes while I put this on." Neva mixes up some terrible-smelling purple solution and spreads it over all the hair that the women have pulled through the cap.

Now, as Crystal's friend appears at the door to learn that Crystal can't come out to play, that the streaking will take another two hours, the women resume another, earlier conversation about a fatal car accident and the state of the deceased's soul at the time of death, meanwhile taking off the bathing cap so that Crystal can have her hair rinsed and the chemical toner applied:

> "What does it look like now?" Crystal asks.
>
> "What did the doctor say?" Neva has pulled off the cap and now she's washing Crystal's hair, kneading the scalp with her knuckles. . . .
>
> "Said there's five stages of emphysema and he's in the last part of number four," Lorene answers.

At this moment, at the mention of "the doctor," the reader is poised at the juncture between the kitchen and the on-screen world of the soap opera. Are we, we ask, in Meadville, or in the kitchen? The two are conjoined in an instant of confusion—until Crystal deciphers the referent, and realizes that it is her own father who has been diagnosed as the victim of advanced emphysema. She cries out through a towel, "but the women are watching TV because Karen is telling Gregory everything." In the next moment both the TV character and Crystal are crying as, on-screen, the issue of caring for children connects the soap opera with Crystal's life.

It is self-evident that the novelist has been deliberate in her choice of the televised scene, one in which two issues bear directly on Crystal's life, the first of a teenage girl estranged from adults, and the second the question of who cares for the children in parental absence. Both can be brought into the fiction precisely to converge with the protagonist's life.

Much can be said about the notion of flow in this scene. Our principal interest, however, lies in the demonstration of teleconsciousness in a habitat in which the television is more or less "always on." There is no sense here of television as distraction from a primary focus of attention, or as interruption. There is no sense of disjunction between the kitchen characters and the on-screen figures. There is no sense whatsoever that life in the kitchen ever comes to a halt for the duration of the program, or that these characters wish it could or feel that it ought to. Their relation to the on-screen world is interactive, with a fluidity and simultaneity among their conversation, their activities, and the TV program. The linear form of the printed text makes it necessary for the novelist to position television and kitchen life sequentially and alternately. But we understand that, just as one's own realm of thought includes the interior of the mind as well as the exterior environment, so now in the TV age, it encompasses the on-screen TV realm as well. And the text claiming to represent contemporary life must include television in a dynamic, tele-conscious way. Accuracy in verisimilitude demands it. The writer no longer goes on the road in an escape from television, as Kerouac did in the 1950s.

The art critic and poet, John Ashbery, in a short fictional sketch entitled "Description of a Masque," offers a description of the postmodern sensibility in terms capturing the quality of the new media-age consciousness. His comments are especially incisive in recognition that a sense of setting has replaced one of history, that history itself is now one part of a mentality of presentist collage or "setting." This setting, says Ashbery, "would go on evolving eternally, rolling its waves across our vision like an ocean, each one new yet recognizably a part of the same series, which was creation it-self. . . . Episodes from history; prenatal and other early memories from our own solitary, separate pasts . . . calamities or moments of relaxation . . . scenes from movies, plays, operas, television" (27). The multimedia mix is one in which the mental and material environments merge experientially, consciousness itself a combination of interactive elements of which television makes up a substantial part. Settling into a scene from *La Boheme*, Ashbery suddenly interjects "murky scenes from television with a preponderance of excerpts from Jacques Cousteau documentaries with snorkeling figures . . . and a seeming excess of silver bubbles . . . to sweep to the top of the screen, where they vanished. There were old clips from *Lucy*, *Lassie* and *The Waltons*; there was Walter Cronkite bidding us an urgent good evening years ago" (28). All this, says Ashbery, makes up the contemporary state of mind.

And it corroborates the state of televisual interaction as the onetime TV babies represent contemporary life in a media environment. These writers may glance back nostalgically to a first-wave TV childhood of family-centered shows, as Susan Minot does in *Monkeys* (1986): "On Sunday nights we have treats and BLTs and get to watch Ted Mack and Ed Sullivan" (11). But most texts represent consciousness simultaneously engaged in two (or more) environments, watching television, talking on the phone, talking to other(s) in the habitat—even reading, as one sketch on a TV-age writer shows: "He flits from medium to medium, reading *Science News* and Hans Christian Andersen and Homer by the light of the television set, the lava lamp that is never off when he is on" (Hall).

Surprisingly enough, the new teleconsciousness was anticipated as early as 1944, when the DuMont Corporation projected the idea of television as environment. Someday soon, its ads promised, "you will be in two places at the same instant. You will be in your living room and at the Presidential Inauguration . . . you'll be in your easy chair and at the opera . . . you'll be with your pipe and slippers and with the team of your Alma Mater as it charges down the field" (*Harper's Magazine*, 188 [March, 1944]: n.p.). The idea of being simultaneously in two places, one of them a TV environment, was a radical suggestion for the time, but one ultimately to become a commonplace and to make its way into contemporary representations of consciousness. Here, to cite another 1980s example, is Henry, the protagonist in Frederick Barthelme's novel *Second Marriage* (1984) in a domestic moment that deserves attention not only because it shows the contemporary experience of the TV environment but because it goes further to enact the continuous reprioritizing of attention in the habitat of the "always on" TV. This scene shows, in addition, the individual as an active agent in assigning priorities. In this scene, told in the first person, Henry's wife, who is outdoors gardening, refuses an early lunch together and puts him off by suggesting that instead he watch TV or read a book.

"Gee, thanks," I watched her dig for a few more minutes, then went inside. The television was on in the living room. I sat on the couch and watched for a few minutes, a marathon, a lot of ordinary-looking people plodding through some pretty countryside, California I think it was.

Cindy called. "Hi. How are you?" she said.

"Fine," I said.

"Listen, I just wanted to tell you about this sale at Radio Shack. We were over there. . . . They're closing out all their Intelevision stock. . . . We got . . . games . . . Snafu. Bull Frog."

"Uh-huh." I was watching a pretty young girl with black hair and crimson shorts cross the finish line on television . . .

"I told Duncan we ought to have you guys over to play some of these" . . .

The marathon girl was bent at the waist, her hands on her knees, panting hard. She straightened and started walking in lopsided circles. The camera came in tight on her face. I could see the down on her cheek and the big bubbles of perspiration. "First we have to get [my wife] Theo out of the back yard," I said. "She's out there digging a hole."

"Yeah, I saw her out there . . . planting . . ." . . . "Well, it's great exercise. Digging, I mean. Maybe I'll go out and see what she's up to."

"Call me back if you uncover anything, O.K.? I'll be in here watching this race on television, this marathon."

"Call you back," she said. "You're watching a marathon?"

"I'm not watching watching. I'm just looking at it."

I'm not watching watching. I'm just looking at it. These two sets of terms, immediately apprehensible and familiar, demand to be noticed in opposition to one another. "Just looking" indicates a scanning gaze, a level of attention that affords the individual freedom for other activity carried out simultaneously. "Just" is itself a dismissive term. To be "just" looking is to be engaged minimally. This perceptual mode, as *Second Marriage* and other fictional texts indicate, is the more customary of the two. The very term, "just looking," echoes from its source in retail shopping, in which it fends off the approaching clerk, that agent of commercial transaction, and reclaims for the speaker a personal zone of free space. The individual, whether a browser or a would-be buyer unwilling to disclose his or her intention, dispatches the clerk with exactly that phrase, "just looking." Used here to describe TV watching, the term indicates the self's conscious distancing from televisual environmental immersion. It also bespeaks a free-space zone for the "looker."

"Watching watching," on the other hand, means deep concentration, as Henry and his caller Cindy know, a commitment of time and focus. The term indicates a first-priority involvement in the on-screen environment, an involvement that would necessarily exclude and preclude other demands on attention. Implicitly, in addition, "watching-watching" versus "looking" also suggests a point of etiquette; it would be rude for Henry to seem attentive to his neighbor–caller were he actually committed to the marathon while appearing to hold a full phone conversation with her. He voices the distinction to assure her that he is listening and that he is also willing and able to come visit.

What does Barthelme's scene with its crucial line of dialogue tell us about contemporary television?—that the individual *knows full well* at any given moment whether the TV experience is one of "watching watching" or of "just looking." The juxtaposed terms make explicit what is known but unexpressed, that there are varying degrees of assent to, and participation in, the TV environment, and that the viewer is conscious of them and able to articulate that consciousness. Arlen's Thanksgiving Day fable had argued that Father was captive to the TV world that drew him, virtually helpless, from his holiday pumpkin pie. Given the example from *Second Marriage*, Barthelme would argue on the contrary that Father made a conscious decision to escape the annual family banalities in favor of the far more interesting TV football game. It goes without saying that this particular scene of Barthelme's, compressed here to one-third its length in the novel, could also be read for the erotic interest of the male gazing, mildly stimulated, at the female marathoner. And it could be read for sociological content, showing the hip Yuppie suburbanites preoccupied with electronic toys. In enacting the contemporary TV experience and giving voice to its hierarchal distinctions, however, the novel reveals much about contemporary teleconsciousness.

And teleconsciousness is not limited to recreational hours or to women's domestic sphere, but extends even to business, with its ethos of efficiency and production and its continuing aura of masculinity. The business executive is our concluding focus here, and to fully appreciate the new state of mind, it is appropriate to look first at a traditional, pre-TV portrait of the executive, this from F. Scott Fitzgerald's account of the day in the life of the movie producer, Monroe Stahr, in *The Last Tycoon* (1941). Stahr is the quintessential successful modern executive, decisive, powerful, efficient. His command of facts is impressive, his judgment in the industry unassailable. The novel shows his typical workday in which he counsels an actor–star, admonishes two directors, plays host to a visiting dignitary, makes budget decisions, visits sets, reviews raw and edited film footage. The day involves taking Dictograph messages and a number of phone calls, as well as face-to-face discussions. There is no leisure, for "minutes were precious" (51).

Yet Fitzgerald is very careful to present these events in linear sequence. They follow one by one, and even the inevitable interruptions ("Your office, Mr. Stahr." "Oh, yes—oh, hello, Miss Doolan") are presented as items in a linear list. Stahr is never shown engaged simultaneously in more than one activity. Each event merits his full attention. Fitzgerald's presentation of his workday suggests that it is literally unthinkable to suppose that Stahr might engage in more than one locus of attention simultaneously, that someone in his position would even know how to do so. Instead, he engages sequentially: "If he could go from problem to problem, there was a certain rebirth of vitality with each change."

Contrast this with an account, in *U.S. News & World Report*, of a successful corporate business executive of the late 1980s, a DuPont Corporation manager whose third-floor office of his Westfield, New Jersey, home is "equipped with a computer, two telephone lines, a television and a stereo." On weekends at home the executive conducts what he calls "low-intensity work," meaning "he'll sip a beer and keep one eye on a televised ball game while he catches up with his electronic mail" ("The New Organization Man," 49).

The sharp contrast here is one of consciousness. The journalist presumes, without the need for justification or explanation, the power to focus both on the electronic mail (on-screen or via printer) and on the televised image.

Implicitly, the executive could take a phone call at the same time, or turn down the TV sound and play the stereo while keeping the ball game on-screen. The sketch presumes that this divided-and-multiplied consciousness is appropriate for the contemporary, media-age executive. The very fact that this man is identified as a DuPont manager on the rise ensures the propriety of the image. Going public in a national news magazine, he would not take any risk of sending the "wrong" corporate message about his home office. (He does not, for instance, have in the office a small refrigerater that might suggest too much beer on hand. He does not have a VCR, which might suggest a conscious decision to provide entertainment for himself even during "low-intensity" work days.) The technologies in place are those deemed by late-twentieth-century corporate America to be appropriate, and their use argues a new kind of media-age consciousness, not passive, not entranced but multivalent and interactive.

Certification—As Seen on TV

The boy from on television sat playing checkers. He was wearing a velour shirt with a zipper, just like any other kid. But I knew he wasn't. I'd seen him on television.

Mona Simpson, *Anywhere But Here*, 1986

An extraordinary cognitive change has occurred in the TV era, as this moment from a recent novel indicates. Stylistically unremarkable, it seems at first simply to show one character, an aspirant television child actress, describing a boy TV actor she sees in an airport lobby. This girl's world, we understand at once, divides into two categories, the ordinary ("just like any other kid") and the televised, and it is remarkable first because of the traditional categories that do *not* come into play. Recognizing the boy TV actor, the girl does not begin to distinguish his image from his reality. She offers no observation on what the "boy from on television" *really* looks like off-screen playing checkers in street clothes, how different (or similar) he is "in person" from his TV image.

Her categories, instead, are two, the ordinary and the televised. The televised boy may resemble an ordinary boy, wear identical clothing and play the same childhood games, in this case checkers. Simply by virtue of having been televised, however—"as seen on television"—he is not, and cannot be, a figure from the mundane, quotidian world. He has attained some other, different status, which the observer understands at a glance. Perceptually, she does not focus on the gap between the traditional inversions, simulation and authenticity, or between appearance and reality. She operates, instead, from a division between the ordinary and the televised.

This fictional vignette indicates a momentous cognitive change brought about by television, and one that demands attention. Up to this point, we have seen the TV environment take shape in discourses drawing on the ideology and iconography of the American past, for instance the colonial hearth and the concept of individualism. And we have observed the discourses that represent social accommodation to the TV receiver, the hostess snacks and guest behavior, and so forth, together with the messages conveyed by the television lounge chairs and the screen. Having also examined the representations of the viewer's body and mind, one finds nonetheless an additional issue focused on the on-screen world itself. Numerous texts argue that an ontological change has occurred in the TV age, that a great divide exists in representations of those who are—or are not—televised. At issue are status, privilege, the very certification of human experience in the TV environment.

The medium has introduced a new set of primary states of being into a culture long ambivalent about the distinctions between authenticity and fakery. No longer is it solely a question of the relation between the real and the imitation, although that conundrum persists as "the contemporary rage for the factitious exists alongside of the contrary appetite for the authentic" (Orvell, xxiv). But from the late 1940s, certain marketplace and technical developments in the acculturation of television have brought into existence a new set of primary ontological categories. When the semiotician, Umberto Eco, traverses the United States to decode meanings signified in popular culture, including wax museums, former movie mogul mansions, motels, trained animal shows, theme parks, and art museums, he finds that "there is an America of furious hyperreality, which is not of Pop art, of Mickey Mouse, or of Hollywood movies . . . [but that] creates somehow a network of references and influences that finally spread also to the products of high culture and the entertainment industry." Eco, like other analysts of postmodernism, designates this cultural complex of artifacts *hyperreality* and concludes, "The American imagination demands the real thing and, to attain it, must fabricate the absolute fake" (7, 8).

It sounds paradoxical, at least oxymoronic, the idea of attaining authenticity by constructing a simulation. And yet this is precisely what happens in the TV era, when the on-screen image achieves a status traditionally accorded the realm of the authentic. The "boy from on television," as the fictional moment suggests, is special, privileged over and above the ordinary. Hyperreal, he is not to be confused with an "ordinary kid." He is the real thing, supra real, precisely because he is an on-screen simulation.

How has this TV-era ontological change come about? How has the world come to be divided between the ordinary and the televised, with the latter accorded privileged status? Certain critical caveats show that this distinction was not inevitable. Early TV-era texts of the late 1940s–1950s focused on the threat of television, with its awesome powers of surveillance, invading personal privacy. As one critic of 1949 cautioned, "This sorcerer's dream, this ability to make men and women suddenly visible afar without their knowledge, has

Mattel Corporation toy marketed on basis of having been "seen on TV."

turned out to be one of television's most beguiling and worrisome powers." The critic went on to propose plausible (and socially conventional) examples of ordinary citizens exposed on camera in compromising situations—the adulterous husband in public with his mistress, the young woman adjusting her underwear, the sports fan caught on-camera in a burst of profanity. "The man or woman who attends a public function probably surrenders . . . his right to be left alone" (Yoder, 130–31). This kind of warning, together with cartoon humor on the anxiety about surveillance, and the successful *Candid Camera* program whose appeal Andrew Ross has located in the link between voyeurism and surveillance—all show that it was not a foregone conclusion that being "seen on TV" would be perceived as advantageous, suggesting instead that individuals needed to be on guard against television prying into and exposing the personal realm of privacy (Ross, 106–9).

But advertising, not surprisingly, is deeply implicated in the cognitive change. From the earliest days of network commercial broadcasting, television has been a showcase for commodities whose salient selling point was their on

Life-size television

The largest direct-view screen made. A picture unsurpassed in clarity and brilliance. New sensitivity and freedom from interference. Cabinets of distinguished design. See Du Mont — and see television at its finest.

The Du Mont Bradford — Life-size direct-view screen, 203 square inches, on a 19-inch Du Mont tube, FM radio. Record player for the new 45 RPM records. One of a complete line of television receivers.

DU MONT *First with the finest in Television*

1949 DuMont ad in which on-screen image is vivid ("life-size") while viewers are mere outline sketches.

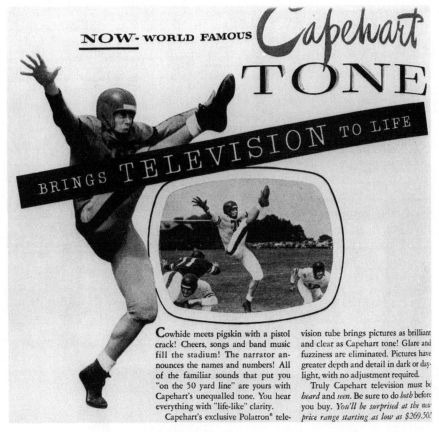

NOW- WORLD FAMOUS *Capehart* TONE BRINGS TELEVISION TO LIFE

Cowhide meets pigskin with a pistol crack! Cheers, songs and band music fill the stadium! The narrator announces the names and numbers! All of the familiar sounds that put you "on the 50 yard line" are yours with Capehart's unequalled tone. You hear everything with "life-like" clarity.

Capehart's exclusive Polatron® television tube brings pictures as brilliant and clear as Capehart tone! Glare and fuzziness are eliminated. Pictures have greater depth and detail in dark or daylight, with no adjustment required.

Truly Capehart television must be *heard* and *seen*. Be sure to do *both* before you buy. *You'll be surprised at the new price range starting as low as $269.50!*

1949 Capehart Corporation ad ambiguously promising to bring "television to life."

screen simulation. "As seen on TV" has been a mark and stamp from the early 1950s, appearing in the United States on products from toys to kitchenware to clothing. Over time, the very phrase has become a media seal of approval, which, from the vantage point of advertisers and manufacturers' increases marketability solely from the TV connection. Acquire these special products, the slogan says, and the on-screen images become three-dimensional and tangible. Objects from the simulated realm of television come to you, the consumer, and television itself with its stamp of legitimacy and glamor can be yours in a commercial transaction. Should you have missed seeing the particular product on-screen, the very phrase, "as seen on TV," on a tag or sticker in a store provides assurance of the correctness of selection, and affirms the national status of the product. The purchase of objects seen on television

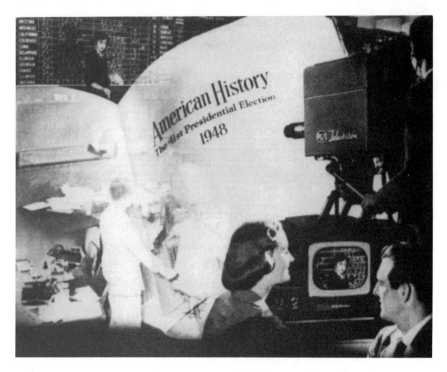

1948 RCA ad emphasizing that historical processes are on view on the TV screen.

means, above all, the certification of one's judgment and the rectitude of personal choice.

Yet the cognitive shift is evidently more directly attributable to another quarter of advertising, beginning with the late-1940s magazine ads that identified the on-screen image itself with life and with contemporary history. RCA invited viewers to "see history in the making—on television," while the Dumont Corporation proclaimed that the TV-era child can "see [the world] as clearly as though she were there, because she is seeing it on the big, clear, *direct-view* screen of a DuMont receiver." Phrases like "brings television to life," "life-size television," "looks alive–sounds alive" proliferated in advertising copy for the Capehart, Dumont, Magnavox, and other TV set manufacturers.

Of course, the black-and-white poor resolution image of these earlier years of television made such claims laughable. But advertisements promising to "put you on the 50-yard line" or to make you "part of the drama and the fun of the world your television brings you," indicated the direction these TV ads would continue to take in persuading the viewer that the more compelling

realm was "the world your television brings you." Tracking these advertisements, one sees the marketplace construction of two separate realms, the one
on-screen and the other off. Implicitly, the particular televised program does
not matter, since the broadcast images per se are by definition better, more
vital, active, centered in world affairs (*New Yorker*, [Oct. 29, 1949]: 105; [Aug.
13, 1949]: 37; [Feb. 18, 1956]: 51; [Oct. 30, 1948]: 35; [Nov. 27, 1948]: 45).

As for the quality of off-screen life, even at its most intense it is implicitly
mundane, removed from centers of action, pallid. The TV realm is the only
desirable place to be. And one notices that the graphics even in these late
1940s–early 1950s ads privilege the on-screen realm, which appears in
photographic high-resolution quality images. The football player, the singer,
and the comedian Art Linkletter all appear solidly, sculpturally three-
dimensional, while the viewers are only sketched in or shadowed in a darkened
living room, or else absent altogether.

Advertisements for color television in the mid-1960s only continued to
reinforce and intensify the division between the two realms, the on-screen and
the off. "So real you'll think you are there," says one 1966 ad for RCA, the
"Color TV Pioneer" (*New Yorker*, Sept. 24, 1966, pp. 120–21). In this
particular ad, we know we're watching the final hole of a golf tournament
because the late-afternoon shadows have fallen across the putting green. The

RCA ad, 1966.

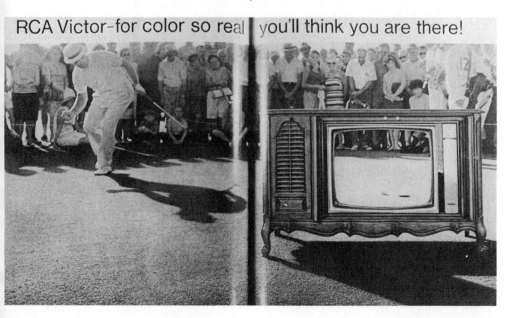

RCA Victor-for color so real you'll think you are there!

crowd and the championship golfer himself are darkened in the failing light. But a huge RCA TV sits squarely on the green, and on its screen we see with stark clarity and in brilliant color the ball rolling toward the hole, inches from dropping in—purely for the benefit of the TV viewer. This RCA color ad, like those of the black-and-white years, enshrines the on-screen image. The copy text may literally say "you'll think you are there," but the graphics say that "there" is not the golf course or for that matter any playing field or environment except the screen itself. Having constructed a spurious competition between the real-world football field and the real-world putting green on the one hand, and the television screen on the other, the ads make certain that the realm of the TV screen wins. And this message has been communicated consistently for over forty years, with the television advertisers continuously enjoining the public to divide the world into two environments, on-screen and off-screen, and to prefer the simulacrum, all that which is "seen on TV."

As our fictional vignette about "the boy from on television" shows, moreover, the meaning of being "seen on TV" has broken the confinement of commerce and reached more deeply into human experience in the United States in this TV age. Assorted texts suggest that the term encompasses a wide range of human concerns. If the on-screen world is *the* place to be, then inevitably, a complex of human concerns—identity, experience, locale, action—these, too, must migrate to the TV screen. Inevitably, the status of these human phenomena will rest on their being "seen on TV." In the 1989 movie, *Batman*, for example, the villainous Joker, played by Jack Nicholson, punches out a television screen in jealous anger after seeing that it is Batman (Michael Keaton), not he, who gets the TV coverage on the evening news after a spectacular city shootout that the Joker has staged for his public inauguration as criminal mastermind. The movie makes clear that in the TV era it is not the contention of evil against goodness that forms the basis of the Joker's rivalry, as it had in the 1940s comic books in which the two characters originated. Instead, this late-1980s film focuses on the competition for media coverage—for being seen on TV.

Although the two film figures are campish, cartoon extremes, the central issue concerns validation by television and has widespread recent precedent in American culture. Regular appearances on TV have become one important credential of major characters in novels, for instance, in Norman Mailer's *An American Dream* (1964), whose hero, Stephen Rojack, is "a personality on television" (15), as well as a former U.S. congressman and author, just as Nora Ephron's Rachel Samstat, the protagonist of *Heartburn* (1983), has her own cooking series on public TV as well as credit for several cookbooks. The television career on-screen seems necessary to complete the character as a fully contemporary figure. A TV show is a credential of modern life. And this TV validation extends into current events. The news journalist Daniel Schorr recounts that in 1977 a deranged man took a hostage, then "led him to the police and the camera crews and yelled, 'Get those cameras on! Get those goddam cameras on! I'm a goddam national hero!'" (164). As such an incident suggests, the importance of being "seen on TV" has reached deep into popular culture.

Television becomes, in this sense, the dominant technology that changes our relation to the natural world. We experience events in its terms, begin to live in reference to it and ratify experience in terms of its on-screen simulation. We live *through* it. Schorr, reflecting on the ontology of the on-screen, proposes a TV-era replacement of Descarte's dictum, "I think, therefore I am," which now becomes, "I am seen, therefore I am." In this sense the on-screen world ratifies existence and becomes, moreover, an environment. The term, televisionland, becomes a world in itself and not an ironic and hip comic designation at which all but the most credulous are invited to wink. To cite just one more example—in 1989, the Hallmark Company sold a greeting card in its "To Kids with Love" series, featuring a televised parental bear holding out a snapshot of an ursine offspring and extolling the young bear's virtues—all on-screen. The card and accompanying message say that the best, most appropriate venue for expression of parental pride is the TV screen. Both bears, parent and "kid," are "seen on TV," and the TV on-screen simulation verifies the authenticity of parental love.

From the 1950s onward, and well before the camcorder and VCR put large segments of the U.S. population on the television screen, various texts were revealing the ways in which being broadcast—"as seen on TV"—began to constitute a new kind of ontological state in which the self, its place, its actions are ratified and validated. In fact, TV-age texts mark the change in which the necessary point-of-view is no longer that of the audience but of the on-screen participant viewing oneself on the screen. A pair of examples from fiction can help to indicate the course of this change. In a 1950s murder mystery novel, two killers named Sharkey and Chop argue with their moll over when and how to commit homicide and dispose of the corpse, tailoring their crime plans to the TV program schedule ("It's eight-fifteen now and I wanna be upstairs when Bob Hope comes on. . . . Goddammit it, . . . now I'm gonna miss Bob Hope" [Goodis, 79]).

By the 1980s, however, the murder mystery is more likely to present characters being taped and broadcast, and then watching themselves as on-screen participants "seen on TV." In Scott Turow's *Presumed Innocent* (1987), the narrator, an assistant prosecuting attorney involved in a notorious murder case, approaches a funeral chapel to see "the minicam van of one of the stations, complete with its rooftop radar dish, . . . and a number of report-

Sony Corporation ad, 1970.

ers . . . thrusting microphones at arriving officials" (11). That night, the attorney watches the TV broadcast with his wife: "Without even noticing at first, I become aware that both of us have ceased comment, even motion, and are facing the television set, where the screen has filled with images of today's service. . . . Raymond's car arrives, and the back of my head is briefly shown" (38–39).

Not every murder mystery novel includes television, of course, but a significant shift can be observed in those that do. In the 1950s, the character was presented as a member of the audience; in the 1980s, the character becomes a TV participant, both videotaped and broadcast, and self-reflexive in watching the TV image of the self. The cycle is only complete when he is "seen on TV," even in partial view, and seen by himself as well as others. The self is ratified in the broadcast of its own image.

Thus, the slogan, "as seen on TV," once formerly attaching to purchasable products, has come to imply much more. Still best understood in the United States as a marketing device, the slogan points further, toward the cultural change wrought by television in the introduction of a new set of cognitive and perceptual categories in which what is real or unreal is superceded by what is

1989 Hallmark card in which parental bear appears on TV to express pride in offspring. Reprinted courtesy of Hallmark Cards, Inc.

televised and what is not. "As-seen-on-TV" continues to tag the packages of myriad products, but in the meantime certain American texts suggest the presence of deeper cultural meanings codified in the slogan. As the range of what is seen on TV—persons, groups, places, activities—increasingly defies enumeration, and Andy Warhol's projection of fifteen minutes of fame for everybody makes every individual a candidate for the tag, "as seen on TV"— we need to look more closely into the nature of the new, TV-era ontological categories.

At first, this new status seems merely narcissistic. The classic Greek myth of the youth, Narcissus, mistaking his own reflection for that of another and becoming enamored of his own image mirrored in a pool has haunted the TV age in which Narcissus has seemed to represent the viewer at his (or her) most self-obsessed, impervious to the outside world except as consumer, all powers of judgment gone, consciousness telescoped to the self-referential. Television in this formulation is the drowning pool. And from Marshall McLuhan to Mark C. Miller, critics have confronted the idea of the viewer as tele-narcissist. McLuhan theorized that the apparent apathy that is thought to be characteristic of electronic-age narcissism ("Narcissus as Narchosis") is really a physiologic defense mechanism against sensory overstimulation (51–56). Miller, invoking terms from George Orwell's 1984, argues that, "guided by TV, we watch ourselves as if already televised, checking ourselves both inwardly and outwardly for any sign of untidiness or gloom, moment by moment as guarded and self-conscious as Winston Smith [Orwell's protagonist] under the scrutiny of the Thought Police" (328).

Yet to be "as seen on TV" is different from the narcissistic. It is not the self-contained, closed system of the self and its reflection that constitutes this state. Instead, central to the televised state, being "seen on TV" is the transposition of the self and its activities into the on-screen environment of television. The transposition itself is crucial. Reflection, which is the narcissist's sine qua non, is subordinated to it. The reflected images essentially give evidence of the transposition onto the TV screen. To be transposed onto television is to be elevated out of the banal realm of the off-screen and repositioned in the privileged on-screen world.

At issue here is the way in which self, place, objects, and experience come to be ratified by television, and this phenomenon has been explored in Walker Percy's *The Moviegoer* (1960), a novel that interests us here because Percy's characters experience the very kind of on-screen status conferred by television as it became the dominant media environment through the 1960s and beyond. In *The Moviegoer*, the protagonist–narrator, an avid movie fan, takes a young woman to a movie in which they watch on-screen scenes of the very New Orleans neighborhood surrounding the movie theatre where they are sitting.

In the darkened interior the couple watch the images of their immediate environment. Once outside after the screening, they look about, and the woman declares the place "certified," the term clearly planted by Percy to permit an observation about the power of media simulation. "Nowadays," he writes, "when a person lives somewhere, in a neighborhood, the place is not certified for him," meaning that the individual will find life there both sad and irremediably empty unless "he sees a movie which shows his very neighborhood," at which point "it becomes possible for him to live, for a time at least, as a person who is Somewhere and not Anywhere" (55).

Percy detests this "certification," in which one's own life is denied validity or forfeited to a meretricious media image. He scorns the privileging of media-generated illusion. Such a spurious life, Percy argues, is the result of alienation from the self, detached and estranged in an age of hegemonic science and technology, a world in which the constitutional pursuit of happiness instead degenerates into "the pursuit of diversion" (*Lost in the Cosmos*, 12).

Percy couples his definition of media-age certification with outright repudiation of its delusion, illusion, alienation—in sum, its unreality. *The Moviegoer* urges its audience to get its moral bearings by rejecting the spurious media world. A person properly claims title to life largely by differentiating the self from those enthralled to the media image. Some twenty years later, in the 1980s, Percy engaged the television generation on these same terms:

> On the Johnny Carson Show, it always happens that when Carson or one of his guests mentions the name of an American city, there is applause from those audience members who live in this city. The applause is of a certain character, startled and immediate, as if the applauders cannot help themselves. (*Lost in the Cosmos*, 27).

The TV audience in the Burbank, California, studio feel so dislocated and detached from their native places, this text argues, that hearing the place named on TV awakens them to the realization that their own place, whether a byword like Chicago or an obscure site like Abita Springs, Louisiana, is real and true. The Burbank TV studio becomes the scene of therapeutic shock treatment for the studio and TV audience. It is a kind of space station from which these televisual aliens can get their space-time coordinates and navigate their way home at last. Percy's subtitle, *The Last Self-Help Book*, suggests that the TV audience in Burbank, experiencing dislocation, can now, perhaps for the first time in their lives, go home with a new understanding of their relation to place. By extension, that studio audience is all America invited to take a lesson about real life and media pseudolife, and enabled at last to choose the real, the valid, the authentic. Their true lives are "Somewhere" after all, while the media images are merely "Anywhere," meaning nowhere. These positions

BIGGER-THAN-LIFE! THAT'S THE GIANT PICTURE YOU GET WITH THE NEW 24-INCH "BAYLOR". EBONY FINISH. MODEL 249512, $299.95.

The big new television thrill from RCA Victor—

BIGGER-THAN-LIFE 24-INCH TV WITH THE "ALL-CLEAR" PICTU

RCA ad, 1955.

are based, as we see, on the premise of two fundamental categories, the real and the unreal. Percy interprets the spontaneous applause of the TV audience on that basis.

But the slogan, "as seen on TV," insinuates itself here, and is not so easily dismissed or even discredited by Percy's moral categories. What if the applause does not indicate a world in which people must struggle against illusion to claim their place in reality? What if, instead, they applaud the naming of their hometowns because those towns, named on television, are now ratified by transposition into the realm of the televised, certified by television—"as seen on TV?" This is to suggest that the TV audience may well operate like the narrator of *Anywhere But Here*, seeing the boy from television and employing the operant cognitive categories of off-screen and on-screen to mark the gulf in values between the televised and the ordinary. In this formulation, the Burbank audience and the viewers who identify with them experience their locales as certified on-screen by Johnny Carson. Their applause assents to, gives credence to, the televised. It acknowledges television's certification of their place, their lives. With marked irony, Nora Ephron notices this very pattern in her novel, *Heartburn*, in which the narrator turns on *Phil Donahue* to see "five lesbians who had chosen the occasion of their appearance . . . to come out of the closet." The narrator imagines them "watching contemptuously as their friends chose mundane occasions like Thanksgiving with Mom and Dad for the big revelation," while they waited "for the big one, Phil himself" (62).

Television's ratification of human experience extends beyond the structure of studio or program, and beyond the rubric of entertainment or even celebrity. The hand-held TV camera has made it possible to be taped and "seen on TV" in virtually any situation, as the highly popular program of 1990, *America's Funniest Home Videos*, demonstrates. And this is precisely the point, that a situation in which individuals may find themselves is only ratified if it is taped on site, then broadcast so that the participants can see themselves.

Don DeLillo examines the ontological implication of this cycle in his darkly comic *White Noise* (1985), in an episode in which a toxic gas leak forces

New Yorker cartoon, 1948. Reprinted by permission of *The New Yorker*.

Tom Wilson, "Ziggy" cartoon, 1987. Reprinted by permission
of United Press Syndicate.

families from their homes into a public shelter. Readers participate in their fear
and anguish—and then in their outrage at being ignored by the media, which
has the authority to demarcate the significant from the inconsequential simply
by broadcasting or not broadcasting an event, act, occasion, and so on. "Look
at us," says the narrator, a college professor named Jack Gladney, "forced out
of our homes, sent streaming into the bitter night, pursued by a toxic cloud,
crammed together in makeshift quarters . . . we'd become part of the public
stuff of media disaster" (146).

Huddled in the shelter, however, the group find that isn't quite true.
Quarantined there for nine days, exposed to the toxic cloud (mortally so, in the
narrator's case), they are nonetheless ignored by network television. A walk-on
character, a man holding aloft a small portable TV whose screen is blank,
voices their media-age plight. "There's nothing on network," he cries, "not a
word, not a picture." The local channel has given the story "fifty-two words by
count," but "no film footage, no live report." He rails on about suffering that is
unredeemed by media reportage:

"Don't those people know what we've been through? We were scared to death. We still are. We left our homes, we drove through blizzards, we saw the cloud. It was a deadly specter, right there above us. Is it possible nobody gives substantial coverage to such a thing? Half a minute, twenty seconds? Are they telling us it was insignificant, it was piddling? Are they so callous. . . . Don't they know it's real? Shouldn't the streets be crawling with cameramen and soundmen and reporters? Shouldn't we be yelling out the window to them, 'Leave us alone, we've been through enough, get out of here with your vile instruments of intrusion.' Do they have to have two hundred dead, rare disaster footage, before they come flocking to a given site in their helicopters and network limos? What exactly has to happen before they stick microphones in our faces and hound us to the doorsteps of our homes, camping out on our lawns, creating the usual media circus? Haven't we earned the right to despise their idiot questions? Look at us in this place . . . like lepers in medieval times. . . . But we look around and see no response from the official organs of the media. . . . Even if there hasn't been great loss of life, don't we deserve some attention for our suffering. . . . Isn't fear news?" (*White Noise*, 161–62)

A hilarious moment, this, and one DeLillo uses to set up a media-age variant on the philosophical problem of the tree falling in the forest with no one present to hear it. In this case, television is absent from the neighborhood scene of the toxic cloud, and still absent in its aftermath—therefore depriving the participants of the certification of their own shared experience.

DeLillo depends on his readers' familiarity with the televisual rituals of disaster, in which the on-camera victims voice their outrage at the intrusion of television crews into moments of private anguish. But actually, DeLillo argues, the victims are willing participants in the standard format of the TV ritual, playing the part assigned to them, that of citizens outraged at the invasion of their privacy. Their outrage is merely dissembled because they eagerly seek television's certification of their experience. More than desiring it, they need it for ratification. The real outrage DeLillo's characters feel in *White Noise* is the absence of the TV camera. Without it, their experience is invalid, their suffering unredeemed. They are not "as seen on TV," but snubbed, ignored, in effect nonexistent. Because the toxic gas episode does not merit televising, its victims have no ontological base. DeLillo's text tells us that the participants can only believe their intense experience is significant if it is televised, and if some of their group view the telecast. The experience can only be certified if the participants see themselves imaged on-screen describing events and interacting with reporters. But network television has judged the death-dealing toxic cloud as too ordinary—and this is the final, crushing insult, that the event and the experience do not merit the transposition from the ordinary into the televised. There will be no on-screen certification, no "as seen on TV."

As examples of this kind proliferate in printed texts, the terms in which they are discussed remain problematic, largely because of the archaic perceptual–moral categories that Percy and his like invoke. These, it goes without saying, force one into contempt for DeLillo's characters in their shelter. They also force one into binary absolutist responses. If one wishes to enlarge the realm of inquiry beyond these unequivocal, absolutist terms, then a different lexicon must come into play.

Until recently, analysts have groped for terms expressive of the new televisual state that they experienced as epistemic dislocation. In 1960, the poet Randall Jarrell observed that people were longing to "be reminded continually of . . . the created world of the Medium—rather than be left at the mercy of actuality" (162–63). By the mid-1960s, one analyst mused that it was no longer clear "what an 'authentic' experience" was, because "direct experience [had] become indistinguishable from the vicarious, the compelling secondhand version" (Wheeler, 22). As of the 1970s, this kind of polar oscillation between the Medium (TV) and actuality, or between the direct and the vicarious, moved a writer to declare, "We are honestly and thoroughly at sea both in fact and fiction," the crucial phrase, "at sea," indicative of experiential fluidity supplanting two previously clear and bounded states of being ("Sedulus," 32). All such commentators conveyed their sense of unwelcome ambiguity even as they recorded an epistemic change for which they had no terms. In 1977, Daniel Schorr, professionally committed as a broadcast journalist to the differentiation of news from entertainment, wrote that "the unreal filters from the 'tube' into the home along with the real, and gradually the thin line between reality and fantasy becomes eroded in the minds of many" (156). Such comments acknowledge the process of epistemic change. They record the struggle to describe the new televisual state of being in a postmodern world—but then fall back on pejoratives from older, familiar cognitive categories in which television is "the secondhand version" of experience or the "national seance," or merely "the media's configurations" of events (Schorr, 156; "Sedulus," 32). At base, analyses like these invariably reaffirm the existence of the "real world" locked in combat against the "made-up world." When "reality loses," the analysts' dismay is implicitly a summons to reform, meaning renewed battle for "real world" victory and reaffirmation of the boundaries between the traditional categories (Schorr, 106). And thus these postulates, premised on the division between real and unreal, persist in the decades when "live" television is routinely a videotape, when instant replays proliferate, when forms of reenactment, simulation, recreation, and

Journalist Daniel Schorr on-screen and off-screen. *Esquire* magazine, 1977.

docudrama are commonplace on television, when within television itself the lines between genres and formats blur.

The issue is not simply whether one participates in a TV-era jeremiad, lamenting change as sociocultural degradation. Nor is it coercion into contempt for the likes of DeLillo's media-deprived refugees. The point, rather, concerns the terms in which the phenomenon of the on-screen per se and of being "seen on TV" can be cast. To escape the binary opposition with its absolutist morality, one must return to the initial phase of this discussion and ask again, What does it mean to be televised, "as seen on TV?" Here is a vignette recounted by Schorr from his own days working in network television:

> In May, 1972, CBS News was videotaping a preconvention Democratic platform hearing on foreign policy in New York City. Several persons in the audience wandered from their seats to where I was working and peered in fascination at the witness, Averell Harriman, on our tiny black-and-white monitor. When I pointed out that there was nothing on the monitor they could not see in the flesh only twenty feet away, a middle-aged man replied, "Yes, but it's more interesting when you see it on television." (105)

The question becomes, What does "more interesting" mean? Given the choice between the off-screen and on-screen experience, the very kinds of experience offered at baseball and football stadiums, basketball sportsplexes, racetracks, and the like, these people, such as the fans, can choose the latter without evident anxiety about the vicarious, about feeling all "at sea" or anxious about the erosion of their powers of discernment. The on-screen world is intrinsically more compelling in a way not expressed in the traditional language of antitheses. And the customary designative vocabulary, starting with media and image, referentially fails to convey a sense of this new state, even if we decide to invert the now inoperant term, reality, in a distanced, self-conscious, and ironic version, "reality." (For "reality" only mires us in self-contradiction, because the real and the televisual "real" are based in the presumed distinction between the illusory and the authentic.) To add the prefix, "neo," is to default to mere chronological progression.

At this point we necessarily revert to Eco's paradox, which says that the "'completely real' becomes identified with the 'completely fake'" (7). This is the realm also addressed by the French anthropologist, Jean Baudrillard, who argues that categories like those Percy and his cohorts invoke or imply—real/unreal, authenticity/imitation, first-hand/vicarious, actual/illusory—are now superannuated, persisting in our discourse because we invoke them from unexamined convictions inherited from a pervious age. These categories, Baudrillard says, no longer pertain to the condition of things in a media age. Even such terms as imitation or reduplication are now beside the point, he says, are in fact invalid. Baudrillard writes, "It is rather a question of substituting signs of the real for the real itself . . . simulation is . . . the generation of models of a real without origin or territory: a hyperreal" (166). And the contemporary yearning is to participate in that hyperreal, in this case in the on-screen.

DeLillo, once again, has been exploring this state in relation to television and intense human experience. *Libra* (1988), his novel about Lee Harvey Oswald and the events surrounding the Kennedy assassination, suggests, first, that erotics, including women's fantasy of the ideal male, is an on-screen TV experience of the hyperreal. DeLillo shows the fantasy life of Marina Oswald, Lee's Russian wife who is as obsessed with John F. Kennedy as is her assassin–husband. "It's as though he floats over the landscape at night, entering the act of love between husbands and wives. He floats through television screens into bedrooms at night" (324). The JFK–simulacrum is here, and DeLillo suggests that erotics is the TV simulation.

Not that DeLillo is the first to explore the issue of TV erotics. A 1955 story in *The Saturday Evening Post*, "Her Shadow Love," describes a Brooklyn housewife and mother of two besotted by Liberace, the TV pianist whose trademark candelabra and white tie help lure her into fantasy eons away from life with her workman–husband. Each night at six thirty, she abandons the household for televisual romance until at length the frustrated husband wins her back by smashing the television tube with a hockey stick and presenting his wife a dozen roses, a sign that he, like Liberace, can be a gentleman (Heuman). Yet this 1955 story presupposes that the television simulation is merely an illusion that can be smashed in order that real life can be resumed. Illusion can be shattered, reality reclaimed. By the 1960s, however, illusion has yielded to the hyperreal, which is not to be banished in combat against the real.

In fact, television becomes a life-and-death simulacrum as DeLillo goes on to show Lee Harvey Oswald watching himself die on TV in the Dallas, Texas, jail: "He could see himself shot as the camera caught it. Through the pain he watched TV. . . . The only thing left was the mocking pain, the picture of the twisted face on TV. . . . He watched in a darkish room, someone's TV den. . . . He was in pain. He knew what it meant to be in pain. All you had to do was see TV. Arm over his chest, mouth in a knowing oh" (439–40). DeLillo's reader of a certain age will recognize the indelible image of Oswald at the moment when he is shot by Jack Ruby ("mouth in a knowing oh"). But Oswald, watching himself die, becomes the on-screen simulacrum. Losing consciousness, dematerialized into a focus of pure pain, he understands his state in reference to TV watching. He is the self ensconced in, and de-materialized by, the darkened TV den. His memory's fund of TV images of pain informs him that his on-screen face is certifiably one of pain. Television structures and certifies his own death, and records it with a validity equal to that of his own neural system. (And we are mindful of McLuhan's argument that television, like all communications media, is an extension of the human nervous system.) As novelist, DeLillo marks out the postmodern territory in which even life–death human experience is understood in terms of televisual simulacra. The simulacrum is privileged in status, and it assigns meaning to the off-screen world.

Visual artists, too, have been examining the ways in which the TV simulacrum, the image "seen on TV," assigns meaning to a context, both dominating and simultaneously reducing the importance of that very context. In the comic strip, *Doonesbury*, Garry Trudeau's character, Mike, watches a television interview with John Y. Brown, the then-candidate for governor of Kentucky, and his wife, the former Miss America and TV personality, Phyllis George. In one panel of the strip, Brown is heard saying to the interviewer, "I told my ex-wife at the time of my divorce that the only woman who really appealed to me 'in all the categories' was Phyllis George. . . . We'd never met. But I'd seen her on TV!" (*The People's Doonesbury*, n.p.).

Here Trudeau presents the Brown–George pair as another hyperreal version of the much-televised Kennedys ("John and I have always been so fascinated by Jack and Jackie," says the on-screen Phyllis). In the comic strip, Mike stands for the reader–viewer, hearing the TV language about "all the categories," hearing that this gubernatorial candidate believes the on-screen image of a woman to be sufficient (even preferable to the actual individual?), and that Brown and George, presented in *Doonesbury* as on-screen images, ratify their life as a couple in terms of other on-screen images, including their own (*The People's Doonesbury*, n.p.). Panel by panel, then, Trudeau is as analytically suggestive about television as the novelist and theorist of the medium, DeLillo or Baudrillard. Yet one does feel that the various cluttered interiors of the kind in which *Doonesbury*'s Mike sits watching television represent worlds of the real in Trudeau, that he is loath to give up the distinction even as those interiors prove to be so emphatically mundane in contrast to the hyped world of the on-screen simulation.

Other artists, however, immerse themselves in television's hyperreal to enact it and comment critically on it. The Nashville-based photographer, Meryl Truett, presents a series of photographs in which the televisual hyperreal is privileged over all other environments. Meryl Truett composes her series of black-and-white photographs by selecting as context a natural landscape (a pasture, a cornfield, the woods), then setting a TV receiver in it and positioning a familiar media image on the TV screen in the chromatics familiar to most viewers of broadcast television. In some ways, her techniques are like those of the commercial photographer in the RCA color ad showing the TV set on the putting green. In each photograph, the on-screen image is both preeminent, dominating the scene in which the TV console is positioned, and also assigning meaning to the entire landscape. In "Duke and Dogies," one's eye moves first to the TV image of the actor John Wayne (the "Duke") in the role of Rooster Cogburn in the movie *True Grit* (from Charles Portis's novel). Several levels of simulation are present here, as the actor plays a movie part developed from a novel and later shown on television, then is transposed by the photographer from one photographic medium into another. But above all, the television is paramount and assigns meaning to the rest of the photograph, and to the world it represents. The on-screen image of Wayne certifies the cattle in the photograph as icons of the American West (even though it is not at all clear that the photograph was shot in the geographic West).

Meryl Truett makes her most trenchant statement in "Maize, What You Call Corn," in which the TV image is a close-up from a Mazola Corn Oil TV ad of the early 1980s. In the photograph, the bright yellow ears of the TV simulation dominate the background of the cornfield. The hyperreal takes precedence over the off-screen actuality. And here Truett parts company from the commercial photographer, because she so severely drabs her photographic contexts, so bleeds them of textural richness that we understand the degree to which she feels that photography itself is robbed of representational power in the TV era. The commercial televised image, not the documentary photo of the cornfield (nor the referential actual cornfield), becomes the focal experience of corn. The eye cannot move elsewhere in the photograph for long but must return to the TV simulation, which is the ratification of corn itself. (And in this regard, an anecdote is appropriate here. In Chicago, 1989, a woman told me of her son who works on an Alaskan cruise ship, who reports that when the ship sails past a glacier—this in good weather—the cruise passengers are apt to stay below in their cabins, experiencing the glacier by watching its televised image, which is relayed to them via a hand-held camera panning the view from the ship's deck. It is unlikely that these cruise passengers are too lazy to go on deck and probable instead that they have learned to perceive the hyperreal televised image as preeminent.)

Are contemporary writers and artists committed solely to record, and therefore to ratify, the televisual hyperreal? Let us consider the following televisual moments. In a journalist's memoir of events of 1968, there is this report from the Democratic Party convention of that year, when Senator Hubert Humphrey's presidential nomination became official and a film crew from *60 Minutes* recorded the moment: "When the face of [Humphrey's] wife, Murial, appeared on the television in his Hilton suite, Humphrey jumped up

Nashville-based art photographer, Meryl Truett, enacts the hyperreal in "Dukes and Dogies." Reprinted by permission of the artist.

"Maize, What You Call Corn." The title from an advertisement for Mazola oil. By photographer Meryl Truett, Nashville, Tennessee. Reprinted by permission of the artist.

to kiss her electronic image" (Kaiser, 241). DeLillo also presents such a kiss, in *White Noise* in which the family unexpectedly come upon a community cable channel on which their wife–mother is conducting an exercise class. The narrator sees his wife in black and white as "flat, distanced, sealed off, timeless," but their young son, a toddler, "watched his mother, spoke to her in half-words . . . approached the set and touched her body, leaving a hand-print on the dusty surface of the screen" (104–5). Then there is Meg Wolitzer's novel, *This Is Your Life* (1988), its title taken from a popular TV series of the 1950s–1960s, in which two young girls in their New York City apartment watch their TV comedian mother telecast from the West Coast:

> The theme music began, and . . . then mother was brought onstage. Sitting between Johnny and Ed, with the skyline tableau stretched out behind her, she gestured broadly and flooded the entire screen. In that moment the men disappeared, were swallowed up, and even the skyline was eclipsed. . . . *My mother the moon*, Opal thought. *My mother the explosion*. Opal could not take her eyes off her mother. She was madly in love with her, as was half the country. (7)

There is a marked uneasiness about the simulacrum in all these texts. The girls in the New York apartment face their mother years later in person when she has a heart attack, and they all undergo an intense emotional crisis. The journalist, recounting the screen kiss by Senator Humphrey, exponent of "the politics of joy," notes that Humphrey is denounced as "dumb" by a former fellow senator and presidential candidate, Eugene McCarthy (a judgment to which the journalist assents). And DeLillo's child character talking to his on-screen mother is emphatically too young to know better. Even Mona Simpson's narrator, so respectful of "the boy from on television" ultimately becomes a TV performer herself and describes the squalor and boredom of taping sessions, thus exposing their behind-the-scenes reality for herself and for us readers too. The authors here are at pains to demystify and to delegitimate the very simulations they have inscribed.

And Bobbie Ann Mason's fictional characters, for whom television is an important source of knowledge, and who find established officials to be sinkholes of lies and evasions, respond with anxiety to the on-screen simulation to which they legitimately turn for epistemic uses. In *In Country*, one Mason protagonist, Samantha ("Sam") Hughes, a teen in quest of family knowledge and identity, recalls a made-for-TV movie about Vietnam, in which soldiers marched "through a field of corn. The tassels were outlined against the clear blue sky, and the corn looked ready to pick":

> "It surprised her that corn grew in Vietnam. . . . They certainly had corn in Mexico because corn was an Indian plant. Maize. The woman in the Mazola

commercial. It bothered her that it was so hard to find out the truth. Did corn actually grow in Vietnam?" (69–70)

As readers, we participate in Sam's anxiety about the authentic, the valid, the real and true. (And we recall Truett's photograph, which also comments uneasily on the ways in which that same Mazola oil commercial commands the natural environment, damps it down into a sepia background. There is a social statement critical of the simulacra in her photograph, as there is implicitly in Mason's fictional text.)

In all these commentators, there is resistance to the on-screen simulacra. The writers, reflecting their contemporary culture, manifest this postmodern sensibility, yet express reservations about it. They are insiders in the hyperrealist mode, exploiting its characteristics in their several art forms. Unlike Percy, however, they do not position themselves apart from the hyperreal, evidently understanding that there is no legitimate separate place, no valid high moral ground for refugees from the hyperreal. Unable to subscribe to the traditional bifurcations of the factitious and the authentic, of the illusory and the real, they nonetheless reveal their uneasiness about the world of the simulacra and hyperreal. They are resistant insiders. Even the expositors of this new postmodern world resist it, Baudrillard in scathing irony, Eco in advocacy of guerilla warfare in "the battle of man against the technological universe of communication," particularly television (142). Suspicious of the hyperreal, the condition of their postmodern moment, these writers and artists struggle to define and yet simultaneously expose it, working in ambivalence as they demonstrate what it means to be "seen on TV."

Videoportraits and Authority

He turns on the television and lies in bed to watch it. . . . The evening news features . . . a word about former President Nixon. . . . There is a picture of the former President. He looks like a lean old mafioso.

Ann Beattie, *Chilly Scenes of Winter*, 1976

That was the question I had about [John F.] Kennedy. I couldn't decide if he was truly heroic, or just one of those simulated heroes of television.

Lawrence Wright, *In the New World*, 1988

These kinds of irreverent lines seem woven into the very fabric of memoir and fiction of the 1970s–1990s, and they can appear quite casual, as if Presidents naturally draw skeptical or scathing remarks whenever their close-up images appear on television. And why not?—satiric tradition thrives on such iconoclasm. The comedian, Lennie Bruce, convulsed night club audiences in the early 1960s with his imitation of Richard Nixon, and the popularity of Vaughan Meader's record album, *The First Family*, a spoof of the Kennedys, was curtailed only by the November, 1963, assassination. And television itself has broadcast successful presidential satire, from Dan Ackroyd's Nixon through Dana Carvey's George Bush on *Saturday Night Live*.

Desultory observations on TV-era chief executives like those quoted above, however, raise an issue that extends well beyond the prerogatives of satire. This is the relation between television and authority. When the on-screen image prompts comparisons of one president to a mafioso and raises questions on whether one of his Oval Office predecessors is an authentic hero or a TV simulation, we see two irreconcilable positions on the perception of authority. This is authority as it is objectified in high-ranking individuals known to the citizenry virtually exclusively by their images on television. The figures on-screen can range from admirals and senators explaining oil spills and banking bailouts to business executives representing their corporation's positions in ads, televised hearings, or news interviews.

But since these particular observations involve presidents, they direct us to a contemporary division on the perception of authority itself. Linger for a moment with those comments on Nixon and Kennedy. Ann Beattie's text presumes that when shown on television, the image of Richard M. Nixon's face and body can let the viewer grasp the flawed, criminal character of the former president. The television image is presumed to disclose his character, to reveal conspiratorial and criminal traits so clearly that he can be likened to a member of the Mafia. The careful observer, the TV watcher, is presumed to be an active interpreter who can ascertain the inner nature of the subject observed.

The second statement, also responsive to TV images of a former president, is explicit about its confusion. Was John F. Kennedy an authentic hero or only a television simulation of one? The viewer feels he cannot answer the question, only pose it. Television is presumed to put forth simulations, and even the most astute viewer cannot distinguish the simulacrum from what might be authentic.

The point is not to label Richard Nixon a criminal nor to argue John F. Kennedy's status as hero. It is, rather, to recognize the irreconcilable perceptual premises underlying each of the two remarks. The first presumes that television can project personal images—more specifically, portraits—that viewers can read to determine the personality and character of the very highest of authorities. This presumption comes out of a world of sociocultural distinctions between truth and falsity, the real and the factitious, the authentic and inauthentic. In this world, to look closely at TV on-screen images of individuals' faces and bodies is to have the capability of gaining accurate information about the inner self of the individual represented.

The second statement, however, extends itself also to the realm of the hyperreal, of the simulation. Television, this statement presumes, proffers images from two worlds, the one in which reality can be separated from falsity (i.e., in which Kennedy could be confirmed as "truly heroic")—but another world too, in which such distinctions must be set aside because they are not valid, having been superseded by congeries of simulations (in which case Kennedy is a "simulated hero"). The real and the hyperreal worlds slide off into one another, and an individual cannot necessarily know which is projected, and therefore, which he or she inhabits. This cognitive division, then, puts presumptions of sociocultural verifiability on the one side, just as it puts presumptions of hyperreality or simulation on the other. The epistemic great divide, in short, exists at the split between differing presumptions on what is knowably real—and what is hyperreal.

Throughout its history, as analysts of television have understood, the medium has been intertwined with various forms of authority, political, economic, gender-based (cf. Fiske). Of course, any major technology, especially in communications, can be expected to be involved with authority, which one writer calls the spiritual dimension of power (Zuboff, 221). One basis of televisual authority, both Marxists and Frankfurt School critics argue, lies in its very history as a private monopoly protected by the captialists' ally, the federal government, and limiting the vast population to the passive role of recipient while the control of TV transmission remained in corporate hands. Only the receivers, mass produced to lower costs, have been accessible to the population at large, and the authority of television from the beginning has clearly been with the powerful networks and stations. Louis Althusser comments that this communications apparatus reproduces capitalist exploitation "by cramming every 'citizen' with daily doses of nationalism, chauvinism, liberalism, moralism, etc., by means of the press, the radio, and television" (73).

Of course, figures of spurious authority abound in television, and the considerable fluidity of well-known televisual figures in multiple roles and contexts means that the actor playing, for instance, an authoritative role as physician or wise father on a weekly series is apt to surface in ads for medicines or frozen desserts, then to reemerge in the guise of "host" in an awards program or televised charity fund-raiser, and to appear on a public affairs program whose announced intention is informational and educational. The aura of authority in the main role of patriarch or doctor carries forward into each new context (see Fiske, 149–78; Mann, 61–72; White, 135).

To call these figures celebrities is to fail to capture the sense in which they themselves become marketplaces of specious commercial authority. Certain actors (Robert Young, Karl Malden, Bill Cosby) have been notable in such roles in recent years, while the 1960s–1970s TV ads for household and personal products featured superhero, cartoon versions of male authority in the Man from Glad, the Marlboro Man, Mr. Clean, the White Tornado. Both the superheroes and the actors carry thespian authority only, that which is confined to their persuasive function. Because the obedience they attempt to elicit is purely commercial (the purchase of manufactured goods or profitable services), they supplant civil–social legitimacy with the persuasiveness of mere salesmanship.

The discourses showing television's subversion, however, of authority go deeper to the presumed exposure of its human subjects in ways never intended by the industry or its technical or on-camera workers. Numerous texts argue that television creates the conditions favorable for the negation of authority. And this negation, even subversion, is of central concern here because it continues ideas on the perception of individuals, ideas rooted in centuries of tradition in the arts, aesthetics, and philosophy. The subversion of authority by television can be seen as only the most recent form of an ethical and artistic problem that reaches back to the Renaissance and is centered in portraiture and deportment.

In the early 1960s, Marshall McLuhan observed that "technically, TV tends to be a close-up medium," and he went on to build a theory, by now well known, of deep viewer involvement on the very premise that the poor quality image was a kind of gestalt demanding intense viewer involvement to fill in the partially disclosed pattern (276). Though technical improvements over the years have sharpened a great deal the TV image (and thus undercut a major tenet of McLuhan's theory), it remains historically the case that the poor resolution image set the close-up shot as the production standard. (The familiar names for these include the knee shot, thigh shot, waist shot, bust shot, head shot, tight head shot.)

More problematic than image quality was the difficulty of camera positioning or blocking during live production in the earlier decades of television. The shot lexicon was adopted from Hollywood, but the lack of portability of the large, bulky early TV cameras, together with the need for intensely bright lighting and the difficulties of concealing microphones and their shadows meant that while the shots themselves were similar to those of Hollywood, the editing process of television was different and needed to rely heavily on the close-up (rather than, say, the cutaway) to motivate the narrative.

The formal category embracing all these TV camera shots is portraiture, the up-close image that is generically presumed to enable and mandate scrutiny. One observes not only the formal properties of line, mass, and shadow, but the attributes of the character and personality of the individual represented. Complaining that television is "burdened to excess with authoritarian characteristics," the German critic Hans Enzensberger nonetheless remarks that "swarms of liars appear on it, but anyone can see from a long way off that they are peddling something" (119). Here, " a long way off" really means close-up and at a glance, since the close-up shot is the television standard.

And this is precisely the point, as numerous texts argue: that television can and does subvert authority, despite attempts by the industry to elicit respect for it in the format of programs and their settings. The technology and the traditions of transmission have positioned viewers in the uniquely critical stance of viewers in a portrait gallery. As a gallery visitor, the fine arts tradition asserts, one is positioned to read the faces and body language of the video-portraits transmitted. And those faces are often "dirty," meaning that no matter how carefully composed or controlled, there remain aspects of individuals' self-presentation that are thought to divulge to the viewing eye information otherwise censored by managers, consultants, editors, and other television workers (Hartley).

Dirty television, especially in close-up shots, has important implications for the perception of authority. To the viewer self-identified as an interpreter of the image, those "dirty" videoportraits inadvertently subvert authority as much as have portraits in oil and charcoal since the Renaissance, when the individualism of human character and motives became appropriate objectives for the portraitist.

As we know, portraiture, including videoportraiture, is part of a long Western tradition in which the body and the face are presumed to convey information about the state of the soul and the character. Aristotle commented that "the soul's passions all seem to be linked with a body, as the body undergoes modifications in their presence" (in Magli, 88). The medieval pseudoscience of physiognomics presumed to ascertain the fundaments of human passion and intellect by a severe subdivisioning of the body into components in which ethical attributes were thought to be located. By the twelfth century, monastic writings, such as those of St. Bernard, emphasized that gesture itself, including facial expression, indicated moral status ("a wicked man winketh with his eyes . . . the strange movement of the body reveals a new disease in the soul") (Schmitt, 137). From the Middle Ages on, courtesy books emphasized the "body's proper deportment" even for children ("the child is not to sink his head between his shoulders, as this shows arrogance") in order that the psyche's incarnation be presented to the world in appropriately dignified and reposeful signification (Vigerello, 151). This conduct book ethos continued into the Enlightenment, found in Sir Joshua Reynolds's *Discourses on Art,* in which Reynolds indicated that "low and vulgar" characters might be superbly rendered, as Hogarth had done in his Gin Lane drawings of the debauched. But Reynolds insisted that the portraitist of genius knew that imagination was better served by the idealization of noble subjects: "he [the portraitist] must sometimes deviate from vulgar and strict historical truth, in pursuing the grandeur of his design" ("Fourth Discourse," 130–47).

Yet Leonardo Da Vinci's often-cited Aristotelian remark that the portrait ought to reveal "the motions of the mind" not only recapitulates a centuries-long tradition but prepares the way for subversive as well as eulogistic portraits, portraits in which unflattering traits reveal themselves to the discerning eye. Thus Titian's sixteenth-century portrait of Jacopo Strada, a man he disliked, and in whose face he shows pettiness, guile, and a quality one art historian calls "a particularly unattractive sort of eagerness" (Pope-Hennessy, 145–47). This kind of iconoclasm continues in the portraits of Durer, El Greco, Goya, Velazquez, and onto the present, for instance in David Hockney's 1968 *American Collectors (Fred and Marcia Weisman)*, which an art critic observes, shows that Marcia Weisman's lopsided smile "echoes the toothy grimace of a Northwest Indian totem [off to one side] and that a dribble of paint has run down her spouse's fist, as though he were crushing something small and warm to a pulp." Hockney, says the critic, "sometimes allowed himself to prod with a needle" (Hughes, 77, 79).

The issue here is not whether certain critics are right or wrong in their *particular* interpretations of Hockney's portrait or even about Titian's *Jacopo Strada*, but rather that they are exercising a prerogative mandated since the fifteenth century, when portraiture by definition opened itself to this kind of critical interpretation. In accordance with the fine arts interpretive tradition, one does not question the critics' right to make evaluative judgments about the personality traits or character of the individuals portrayed. On the contrary, one expects it and as an educated lay person feels some obligation to engage in such personal interpretation. The task for experts and amateurs alike is not completed until one renders judgment, whether as museum goer or gallery visitor, or as reader or student of art books or viewer of projected slides in public lectures—all roles in which the viewer is self-identified as one whose education equips him or her for such critical judgment.

This obligation extends to the videoportrait, which, interestingly, is self-generated, lacking the mediative presence of the artist or art photographer. As television viewer, one does not face Titian's *Signor Strada* or Hockney's *Fred and Marcia Weisman* or, for that matter, the photographer Matthew Brady's *J. Pierpont Morgan*. The complex triad of relations between the viewer, the artist, and the individual represented is thus much simplified. Before the television camera, the subject becomes his or her own portraitist because the image is self-generated. But viewers self-identified as verifiers of the true, the authentic, the real, are not passive at all. They are interpretively active, scanning, reading the image, judging according to several criteria (ethical, moral, aesthetic, ideological). In fact, the videoportrait, like the portrait, mandates evaluation as the viewer's imperative. By implication, interpretable characteristics are embedded in the videoportrait and must be divulged *to* the viewer *by* the viewer, whether the image shown is a type representative of a class or group or profession (e.g., a spokesman for a federal regulatory agency or a medical association), or whether the videoportrait works toward the individualization of the subject referenced for personal uniqueness (e.g., someone having their fifteen minutes of the fame Andy Warhol predicted for everyone in the TV age).

The difficulty of the task is compounded by one crucial distinction separat-

Titian, *Jacopo Strada*.

ing the videoportrait from that rendered in paint or words. The portrait is a formal program codifying the essence of the figure represented "beyond the accident of the moment" and beyond the vagaries of mood, temporary frame of mind, desultory emotion (Steiner, 11). The portrait, in its Aristotelian sense, strives for a "conceptual rather than an immediately descriptive treatment of its subjects," and thus its rejection of "the accidental" is tremendously important. Reynolds made this point in 1771, remarking that the portraitist who intends "to raise and improve his subject" must approach "a general idea" by leaving out "the minute breaks and peculiarities of the face" (145).

President Richard Nixon telecast to the U.S.S.R. public from Moscow studio in 1972, watched by Mrs. Nixon and then-aide, Henry Kissinger. Courtesy the National Archives.

In the videoportrait, however, the accidental is of paramount importance. It is the source of real knowledge on the viewer's part. It is the crack through which the essence of the individual or his or her type can be apprehended. In the videoportrait, the conceptual presentation of self is understood to be a mere mask or facade, a front behind which the real self is concealed. Thus the critic Mark Miller observes that U.S. presidents from Johnson through Carter "kept lunging at us from behind the mask of power" (82). It is precisely the accidental "lunge" or revelation that discloses the presumed reality of the individual. It is the accidental disclosure of the truth that the viewer must watch for vigilantly.

Certain American traditions give special urgency to this scrutiny for "accidental" truth. The first is the American vernacular democratic debunking of those in high positions. The turn-of-the-century muckraking journalists, for instance, like Lincoln Steffens and Ida Tarbell, were popular for exposing the corrupt practices of powerful figures like John D. Rockefeller. And of course their muckraking, as Theodore Roosevelt termed it, continues in the later twentieth century under the name of investigative reporting. This kind of effort flourishes, not only from prurient interests on the part of the public but from democratically based skepticism about those rising to exalted positions "above themselves" in an ideologically egalitarian society.

Simultaneously, a second American tradition, that of the confidence man, sustains the vigilant watch on the videoportrait. The smooth image of the confidence man conceals his tendentious motives, and he has a prominent place in American texts from the Early National fiction of Hugh Henry Brackenridge through Melville's *The Confidence-Man: His Masquerades* and into this century in Sinclair Lewis's portraits of the confidence man as a religious charlaton and political demagogue. The imperative to expose the confidence man as well as the inflated American lend credence to the very expectation that authority per se, both of types and of individuals, is suspect and must be unmasked. That understanding applies to television. On-screen, then, in minutes, even seconds, the televised individual can present a self as contemptible in the eyes of viewers as the most subtly subversive portrait in oils.

The highly successful *60 Minutes* exploits this very quality. True, the interviewers, in their roles as investigative journalists, portray themselves as the authentic authorities. And to be sure, the program regularly presents flattering portraits of figures in public life or the entertainment industry. But for the objects of the *60 Minutes* investigative journalism, it is apt to be a different story. Often figures of authority themselves—in government, business, the professions—they sweat and squirm or turn belligerent in their own defense. They become weekly portraits of negligence, ineptitude, culpability or criminal liability. Viewers expect to see them as such and to observe their lies and evasions. They expect, that is, videoportraits that in effect subvert authority by showing individuals in it to be cowardly, mendacious, or criminal. The program segments are structured on that basis.

Certain commentators as early as the late 1940s foresaw just this possibility and left a record of their views. Television, said one, "rewards patient viewers" with such exposés as "the possibility of . . . observing an admiral taking a swig of bourbon at a football game. . . . And when Eugene Ormandy was caught in the act of popping a cough drop into his mouth while conducting the Philadelphia Orchestra, there was such a to-do that he publicly swore off" (*Fortune*, May, 1948, p. 4). In the early 1960s, the TV critic, Marya Mannes, suggested the viewer might see "some world leader, some mover of men, whose face and manner might be more illuminating of his character than all his printed speeches put together . . . [or view] a deliberation of the United Nations, where the facial expressions of delegates and the tones of their voices tell more than words, and where the patterns of hostility or compromise are clearly woven" (27–28).

Marya Mannes was not alone in this decade. The theatre director Tyrone Guthrie saw television as the "medium which can so exquisitely satirize the vulgarity and dullness which underlie the glitter of Great Occasions." (During the invocation at the 1952 Republican convention, television zoom lenses dismayed viewers by showing delegates sleeping, absent from their seats and "mill[ing] around . . . without any show of respect or reverence for the Almighty" [Culbert, 185–86].) Guthrie foresaw the power of television to "mercilessly expose the ineptitude of great persons," for instance, the "real cardinal who finds that on a great occasion in a real cathedral he has forgotten his glasses or his handkerchief" (97). As for political life, the pundit Rowland Evans observed, "a candidate directly exposed, face to face [via TV], to virtually every single voter, without the filtering effect of the press or his managers or other members of his party, [can reveal] a single weakness [that] can become a fatal defect" (43). Even Walter Cronkite, the long-term CBS journalist and news reader, commented that because of television's "x-ray intimacy" and many Americans' habit of "going to bed with the television set," the medium presents "a generally accurate . . . but sometimes harsh picture of the person it is scrutinizing and the image it relays" (238).

These various writers are not primarily concerned with the relation between television close-ups and authority. Their purposes vary. The critic was trying to lure intellectuals to the TV screen, the producer to define the role of theatre in the TV age, the pundit to describe political liabilities faced by modern candidates. Yet the point all grasped is the way in which the relentless close-up scanning of the television camera can demystify the televised subject in the eyes of the scrutinizing viewer and consequently subvert authority by destroying the distance between the exalted office and the individual holding it.

Indications are that these observations have proved accurate. In 1955, when *U.S. News & World Report* conducted its survey on television's impact, a housewife was asked whether seeing political leaders on television helped her to understand world affairs better than did listening to their voices on radio. She replied, "I think seeing him very often debunks him. If he's good, he's going to come through good, usually. If he's insincere, I think that somehow shows in his face and manner. . . . when you see them, you aren't so easily fooled" ("What TV Is Doing to Home Life," 46). Nearly two decades later, a survey of public responses to the Watergate congressional hearings of the summer of 1973 found that people across the United States, from Massachusetts to Indiana, Iowa to Oregon responded to public figures—that is, to individuals in positions of public authority—in exactly this personalized way. The researcher found that "many people formed their opinions about Watergate on the basis of the face on the screen." Senator Howard Baker of Tennessee was seen as "impressive," while the Attorney General, John Mitchell, was "more of a Godfather than The Godfather," meaning that Mitchell seemed Mafia-like on-screen. President Nixon's assistant, John Ehrlichman, struck viewers as mendacious. "'He's obviously lying,'" said an

Oregon milkman. "'Look at his sweaty brow and the way he bites his lip like a fish'" (Diamond, *Tin Kazoo*, 42–45). One man recalled, "The media blast hit you. . . . the President of the United States got on TV. . . . Well, I got to tell you . . . that you can't respect somebody just because of their position. People are people. I guess that single thing changed me the most: seeing through authority" (Gottlieb, 44). It is no coincidence that "dirty" television, including "dirty" videoportraiture, becomes most vivid during the weeks' long exposure in a congressioanl hearing room, or from repeated appearances before the cameras and microphones in a State Department pressroom. These spaces become the studios in which the videoportraits are made. It hardly needs to be said that presidents restrict press conferences in favor of infrequent set-piece appearances to guard against what the political pollster, Daniel Yankelovich, found to be Richard Nixon's status with the public, that he so pervaded the media, "he became almost a member of everybody's family" (Diamond, *Tin Kazoo*, 44–45).

This degree of personalization of authoritative figures preceded the Nixon presidency. Lawrence Wright, in his memoir, *In the New World: Growing Up with America 1960–1984* (1988), includes this revealing passage: "by the time the Kennedys came onstage we were deep into the television age. . . . We could see into people's lives. We could peek into the Oval Office and see Caroline and John-John crawling under the President's desk. There was Jack talking politics backstage with Bobby (they didn't even notice us, we were invisible). Jackie took us on a tour of the White House and showed us her bedroom." Wright concludes with this revealing statement: "We knew the Kennedys in the same way we knew the Nelsons, the Ricardos, the Cleavers" (31–32). The presidency, in short, was not unlike a situation comedy, returning episodically at short intervals, its characters as familiar as those on the television programs, *The Adventures of Ozzie and Harriet*, *I Love Lucy*, and *Leave it to Beaver*.

It is perhaps not surprising that in 1970 the counterculture journalist, Raymond Mungo, should speak in tones both of familiarity and contempt when he wrote of Walter Cronkite as "cute and lovable" and of President Lyndon Johnson as "the image of a fat Texan . . . on the TV screen," while a fictional character in a novel of the late 1980s observes "Ronald Reagan's wrinkled red face on the TV" (Mungo, 75, 95–96; McMillan, 157). The tradition of the self-subversive videoportrait continues in TV criticism too, as one analyst lists "all the bad performances we've suffered through for years: LBJ, abusing his dogs and exposing his belly; Richard Nixon, hunched and glistening like a cornered toad; Gerald Ford, forever tipping over; Jimmy Carter, with his maudlin twang and interminable kin" (Miller, *Boxed-In*, 82).

These voices provide anecdotal evidence of a single point, that broadcast

television deflates authority—especially patriarchal authority—via the personalization of the videoportrait. Despite the industry's efforts to construct figures
of authority, the repeated and relentless close-up focus of the TV camera
undercuts it. Statesmen, clerics, presidents unwittingly succumb to the
scrutiny of the lens and the viewers' scrutiny of the relayed image. Familiariza-
tion and personalization undercut the most carefully constructed videoportrait.

And the iconography within the portrait is also significant and directly
relevant here. Of course, iconography inserted within the portrait is a
centuries-long tradition by which the artist locates the represented individual
according to social status, vocation, professional position. The convention
includes the patriarch's family who, surrounding him in the portrait, display
his procreative potency and, in their clothing, his affluence. Great maritime
merchants from Renaissance Venice to colonial New England have been
painted against a background of moored ships, which reference their commercial success and sophistication, and commonly portraits of judges feature
iconic volumes on jurisprudence, while those of scientists contain instruments
of calibration and experimentation.

Television's sequencing, however, makes such portrait iconography a
montage, in which compositional elements elude the control available to the
portraitist in paint or words. This is especially the case of TV images of public
figures and news programs. The imaged political authorities are, in effect,
surrounded by other images that often call the values of the portrait into
question. True, President Nixon or the mayor of Chicago might go on camera
in a studio insulated from the street theatre of politics, and the president might
appear on-screen surrounded by ideological signs of educational, family, and
national values (books, framed photographs of his wife and children, the flag).
But the news segments surrounding his appearance can expand—and
subvert—the carefully chosen iconography of the videoportrait. In television,
there is no insulation from the larger environmental context, which also gets
into the "frame." That larger context often works at cross purposes to
undermine his or her authority. Essentially, the montage of iconographic
images can compete with the videoportrait for credibility and legitimacy when
the ideologies of the two are contradictory and irreconcilable.

Televised historical events of the 1960s–1970s show this well, especially
since political action in those decades recurrently took the form of street
demonstrations against racial segregation and the Vietnam War. Continual
TV images of the police clubbing and tear-gassing demonstrators belied the
assurances of public officials, shot in videoportrait close-ups, that order
prevailed and that the war was just and being won. To look at recorded TV
newscasts of those years is to see the ways in which the televised montage of the

Political cartoon by Herblock (Herbert Block), 1981. Courtesy Herbert Block (Herblock).

larger environment subverts carefully prepared videoportraits of political authority.

The example chosen here is a CBS news broadcast of August 27, 1968, one admittedly vivid in that it occurred during the turbulent Democratic Party convention in Chicago, in which the convention hall was a fortress and the surrounding streets and especially Grant Park the scene of riots. The newscast began with the image of Chicago police marching in line with nightsticks raised, as demonstrators fled before them. Through the course of the thirty-minute newscast, the camera showed these three contextual images: first, a clergyman stating that "the defense of [Chicago Mayor Richard Daley's] political manhood is not the [political] issue"; second, a headshot of a U.S. Army sergeant in Vietnam, his expression grim as he described the death of one of his men, while at the borders of the TV shot were the faces of miserable, frightened-looking Vietnamese children; three, the political analyst, Eric Severeid, also in a head shot, lamenting the lack of executive leadership in the United States, calling Senator Hubert Humphrey "too emotional" and Senator George McGovern "too young and obscure." Severeid sensed at the convention, he said, "a yearning for a man larger than life size." Meanwhile, in the midst of this montage, an ad for cigarettes appeared featuring the Marlboro Man on horseback.

Iconography of this kind is disruptive of conventional portraiture. It provides no eulogistic sign system referencing, in this instance, political executive leadership at the municipal or federal levels. Richard Daley and Lyndon Johnson are discredited in the montage of images surrounding them. Two years later, on a May 5, 1970, CBS news broadcast, Severeid would acknowledge that "for years, erosion has been eating at the authority of traditional institutions," including the presidency. (In these years, the slogan, "Question Authority," also appeared on T-shirts and Volkswagen bumpers.) But Severeid's call in the summer of 1968 for "a man larger than life size" was already a plaintive and anachronistic—and futile—cry for restored, univocal political authority in the age of television's videoportrait. Ironically, the one image of a figure "larger than life size" in that broadcast was the Marlboro man, selling cancer and emphysema in the iconography of the American West.

Videoportraits and their montage of iconography might seem to provide a kind of reading program for television viewers self-identified as critics, but the program is problematic. Even putting aside the myriad biases and predilections each individual viewer brings to these images, a set of historical traditions burdens the most vigilant viewer. Reading the videoportrait, like reading the portrait, belongs not only to a centuries-long tradition of interpretation—but of interpretational anxiety about authenticity. If the conduct books and the Renaissance theories of aesthetics indicated the ways in which body language could be spoken and visually "heard," the process was nonetheless known to be fraught with real and potential errors of judgment and, therefore, treacherous. And that same burden descends to the videoportrait.

Here we might return to the example of 60 *Minutes*, this time to notice its construction of specious authority, instanced in the on-screen reporters, the so-called news team in command of the facts, controlled in aggression, penetrating in insight, entirely trustworthy—performing in their roles of investigative journalists. These individuals are skilled at self-presentation, as are the "anchors" on television network news or the "hosts" on morning programs. And they exemplify the problem of apparent authenticity: the iconoclastic videoportrait of "dirty" television comes mixed with countless other TV portraits that are apparently authentic, imaging individuals who are apparently admirable and appealing, having successfully undergone viewers' scrutiny—but sometimes to be shown up as shams.

The *caveat* of this image gap has surfaced periodically, when trusted, authoritative televisual figures prove, off-screen, to be so unlike the TV image that they are revealed to be exactly that, a mere constructed image. This happened in 1987–1988 in the United States as the television evangelists, Jimmy Swaggart and Jim and Tammy Faye Bakker, were exposed in scandals of sex and financial chicanery, and numerous critics have asked how the public could be so long misled about the character of much-televised political authorities from Joseph McCarthy to Ronald Reagan. And television history includes a series of scandals of well-liked figures. In the 1950s, the disk jockey Dick Clark, a popular emcee of the afternoon dance program, *American Bandstand*, was found guilty of bribery in the recording industry "payola" scandal. Similarly, Charles Van Doren, the personally graceful intellectual who was watched in awed admiration on a quiz program for fourteen weeks, was exposed as a fraud who received answers in advance to the questions he appeared to struggle with—successfully, just in the nick of time, under great pressure—on the program *Twenty-One*.

In a medium technically professionalized to present its programming with utmost smoothness, there is much tension on the part of the viewer positioned between the purveyed image of valid and legitimate authority, and the possibility that such images may be fraudulent. This issue is crucial in television, which distinguishes itself from film in large part on the basis of its "truth-telling." Its claim to broadcast *live*, its news and information programming, its documentaries, morning miscellanies, even its on-air magazines all purport to bring the viewer the facts and truth that constitute reality (Antin, 153).

Even in the Renaissance, portrait artists were acutely aware of the possibility of deception by their subjects. Baldassare Castiglione's conduct manual, *The*

Book of the Courtier, commented that "external appearances *often* bear witness to what is within"—often, but not always (D. R. Smith, 9). In the contemporary moment the presidency of Ronald Reagan renewed the anxiety about a fraudulent, deceptive persona in public life. "Reagan's mask and face are as one," says one critic, who asks, "How does Reagan do it?," then suggests that in part the president's surrounding iconography helped sustain the implacable mask, the "united front." The iconography includes a uniformly corporate cabinet, staffers "who look something like an assortment of boiled eggs," and a First Lady "whose frozen presence at her husband's side suggests, paradoxically, that Reagan is a man of passion" (Miller, 83–84). Yet this critic is as uneasy with the persistently stylized presidential videoportrait as he is with the subverted videoportraits of Reagan's predecessors. The political cartoonist and commentator Herbert Block (Herblock) speaks to this very point:

> In the early days of the medium somebody put out the word that the all-seeing eye of TV could detect true character and could tell truth from falsehood. That observation was itself a first-class falsehood. On television you can't always tell whether the person who comes across as real is the one with or without the makeup—or which person took lessons in how to appear natural. (*Through the Looking Glass,* 11)

Block's point speaks to the problem of successful self-concealment. But he, too, qualifies his statement. "You can't always tell," but then sometimes you can. One finds oneself in another state of tension, sometimes resolved in repeated exposure to the same individual until one has an album (or a rogues' gallery) of videoportraits. And Mark Miller has commented astutely on this, arguing that television is an apparently iconoclastic medium, inherently subversive toward any visible authority, indeed annihilating *"all* individuality" (324). Miller sees this double bind for those in the public eye, meaning the television eye—that public figures, aware that television's "searching eye" will subvert their authority, defend themselves with a cultivated smoothness, "giving up that individuality which TV would otherwise discredit" (325).

Yet all such positions as Miller's, and going back to the Renaissance conduct books, are epistemically grounded in the world of the authentic versus inauthentic, real versus unreal, true versus false. One can err in one's interpretive judgment—as the viewing public erred in misperceiving the demagoguery of Joseph McCarthy and, as detractors argue, misperceiving the cohort of grievously deficient TV-age presidents and so continuing to support them. Nearly thirty years ago, Daniel Boorstin, in *The Image* (1961), a title still in use in the college curriculum, disparaged the new age in which mere oversize images supplant heroes, and pseudoevents replace real news.

But Boorstin embraced the comfortable categories of the authentic and the inauthentic. In texts like his and Block's, the presumption remains intact that the subject of one's scrutiny remains knowable. That subject or figure is presumed to be an individual, a person, a human being who is present even if concealed behind one or more masks. He or she can hide, and the interpreter be fooled, but the anxiety on the part of the interpreter is precisely located along the axis of revelation and discernment. This is finally a reassuring angst—reassuring because it does not disturb the ontological bases of a world presumed to exist in binary divisions between the true and false, real and factitious. The interpreter may be thought to be gulled, naive, ingenuous, but these defects or lapses are located within the individual, just as concealment of the truth about the subject is attributed to his or her thespian success. Either way, the ontological order of things is undisturbed.

That order, however, is challenged by certain TV texts and images. "Until [the 1960s]," writes Lawrence Wright, reflecting on the Kennedy presidency, "I believed what I heard on the radio, what I saw on TV, what I read in the papers. But on the edge of the sixties the system cracked like eggs in a bowl." He adds the comment with which this discussion began: "That was the question I had about Kennedy. I couldn't decide if he was truly heroic, or one of those simulated heroes of television" (Wright, 57–58).

The category of television's "simulated heroes" reintroduces the realm of the hyperreal, one that was widely disseminated in the 1986–1987 figure of Max Headroom, the computer simulation of a TV personality (actually, an actor made up to simulate the simulation). But the notion of the videoportrait-as-simulation inserts the epistemic division avoided in every one of the above-quoted anecdotal statements on videoportraits of liars, debunked pols, mafiosi, and so on. This ontological category, manifest in Max Headroom, explicitly reveals the public awareness (and industry concession to that awareness) of human beings not as dissemblers but as broadcast simulacra. Mark Miller concludes that the successful videoportraiture of Ronald Reagan was precisely the success of the simulacrum: "The best way for the modern [Machiavellian] prince to keep his real self hidden is not to have one" (86). True to the hyperreal, this public authority, President of the United States, is by definition literally self-less. There is no person behind the persona or mask. There is no validity to the preposition, "behind." A 1980s short story, "The New Moon Party," by T. Coraghessan Boyle, profiles such a political simulacrum, the incumbent president "who looked as if he'd been stuffed with sand." Boyle's narrator says, "Oh, they painted him up . . . with smeared blusher and plasticized hair . . . and pointed him toward the TV monitors and told him when to laugh or cry or make his voice tremble with righteousness, and they had him recite his usual litany" (118). And the comics artist, Berke Breathed, exploits the hyperreal in his "Opus" strip, in which President Lyndon Johnson

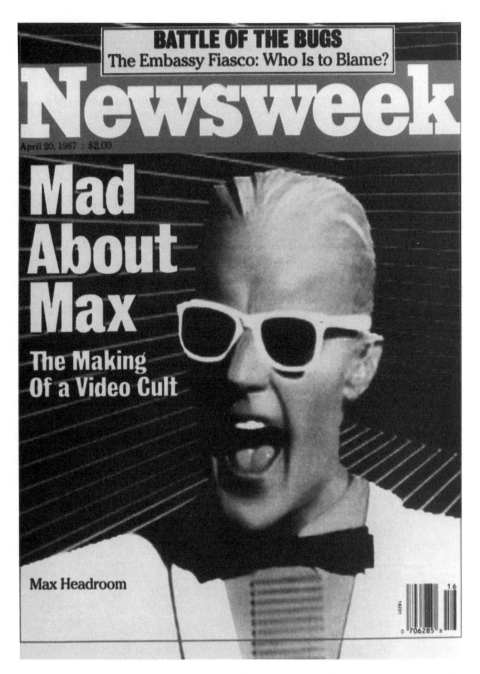

Actor Matt Frewer in the role of "spokesmachine," Max Headroom. From *Newsweek*, April 20, 1987. *Newsweek*, Inc. All rights reserved. Reprinted by permission.

appears on-screen in the company of the notorious atheist, Madalyn Murray O'Hair, and later, in the "corrected" image authorized by the U.S. Federal Elections Committee, is shown with the TV character Bullwinkle from Captain Kangaroo, while John F. Kennedy appears on-screen with Pugsley from "The Addams Family" (106–7). One simulation is replaced by another, a compound of TV simulations vying for supremacy with other simulations.

The short story, the comic, and the visual image of Max Headroom signal the sophistication of a TV-era public poised on the edge of an ontological distinction far more radical in its implications than the storied split over authenticity and the pretense to it. Beyond that centuries'-long division between individuals of good character and those who only mimic it or hide their defects, the figures of the taxidermed president and the computer simulation anchorman indicate another kind of split. This one divides the humane or even human from the simulacral. It voids the very ontology sustained by distinctions between the real and the unreal. But Max Headroom and the plastic president take viewers to an epistemic border most are unprepared to cross. Like the writers and artists currently trafficking uneasily in the hyperreal, TV viewers may be willing to evaluate the televised authority figures on the traditional sliding scale of authenticity and dissembling or lying, yet be unwilling at this time to go onto the next step, to declare human beings as media constructions only, as simulacra. The printed texts of the sort we began with in this discussion on videoportraits say that with the twenty-first century in sight, commercial television and its viewers prefer to be comfortably uncomfortable with the irreverence and the iconoclasm of the videoportrait than to acknowledge entry into an on-screen environment of the hyperreal. At most, viewers may recognize their simultaneous habitation in the two irreconcilable worlds. The first presumes authorities to be three-dimensional figures in representable depths of character and soul, the second as 3-D constructions, simulations, holograms, synthetics—the hyperreal.

Two Cultures and the Battle by the Books

Publishers and authors can only hope . . . that after the novelty of television has worn off, people will again prefer a good book, to the spectacle of two unknown prize-fighters staggering around a ring, or a syrupy-voiced huckster proclaiming the virtues of Dinkelspiel's Deodorant.

Bennett Cerf, *Saturday Review of Literature,* 1948

Fight Prime Time—Read
Bumper Sticker, 1990

Anyone who remembers the popular CBS television quiz program, *What's My Line?,* which aired between 1950 and 1967, will grasp the irony in the first above-quoted statement. Its author, Bennett Cerf, a writer and editor for the Random House publishing firm, became a national television celebrity on that program, in which panelists prominent in the arts and show business tried guessing a contestant's occupation. Seen weekly on television by millions, Cerf became very much involved in the world he initially dismissed for its ridiculous sounding commercial products and pugilistic nobodies.

In 1948, however, Cerf was clearly anxious about the relation of television to the realm with which he was closely identified, that of books and publishing. As his statement shows, he strategically opposed television, which he saw as a medium of nonentity boxers and deodorant, to the book—more pointedly, a "good" book. Not content simply to oppose the two, Cerf stacked his rhetorical deck, essentially eliminating the choice between viewing and reading by making the former absurd. Anyone voluntarily passing an evening watching television as Cerf describes it is contemptible and knows it.

Beyond its particular moment, Cerf's statement is important because it captures a widely and repeatedly expressed conviction on the split existing between two worlds, the one encompassing readers and books, together with other forms of print—and the other containing television and its viewers. From the late 1940s into the 1990s, countless commentators have reinforced that very division, in fact, made it into a barricade separating the realm of print and literacy from that of television. Essentially, Cerf's statement would become the argumentive paradigm sustained over some forty and more years of the TV era in the United States.

At issue here is resistance to technological change by groups perceiving their interests to be imperiled by that change. In this instance, we review a vehement, decades-long survivalist campaign by spokespersons representing the culture of print as they have battled a perceived threat—virtually the threat of annihilation—by television. And this resistance is significant in ways that, for instance, early twentieth-century buggy-whip manufacturers' hostility to the newfangled automobile is not. The culture of print has defined its resistance to the technology of television as a war waged for the survival of the mind. Ray Bradbury's *Fahrenheit 451* (1950) centers precisely on this point. Titled for the temperature at which book paper ignites, it projects a terrifying future of TV walls and of book-burning firemen, a future in which reading is illegal and a single individual crusades to save copies of Plato, Shakespeare, Jefferson, Thoreau, and the Bible from the incinerators to which TV-era totalitarians have consigned them. Bradbury's fable, like other anti-TV public discourse, is a trumpet call to rescue the human mind from a new technological age. His and like-minded discourses have attempted to enlist individuals identified as schoolchildren, educators, parents, or concerned citizens, all summoned to the struggle by texts that define intelligence, maturity, responsibility as traits based in mastery of print forms and of the rejection of television.

The bumper sticker, "Fight Prime Time—Read," reveals the bellicose terms of argument persisting into the contemporary moment. Mutually exclusive, the realms of reading and of TV are locked in combat, and the reader of the bumper sticker is urged to combat television's prime time—the prime cut of the day's hours—by reading. Although communications conglomerates, such as RCA, have owned publishing houses, and although prominent playwrights, notably Paddy Chayefsky, have written TV screenplays, as have a number of the writers blacklisted since the McCarthy era of the early 1950s, still these print-culture crossovers into television have not allayed the anxiety of the print culture. Nor have TV dramatizations of high-culture canonical fiction or short stories or even Shakespeare, for the lines of resistance to the new technology have hardened at the difference per se between the two media, of the screen and of print itself. The campaign of resistance has raged over forty years as the consumer public was urged to gather around the electronic hearth or, as individuals, to sojourn with portables. It has continued as the transposition on-screen of oneself and one's experience became a TV-age imperative. Continually one finds the mutually exclusive categorization of the book and the television, and the a priori presumption of irreconcilable opposition between them.

The lines of contention have been so deeply incised in public discourse that the antipathy between television and the world of print has been taken to be the expression of the truth about the natural order. In 1955, for instance, *U.S. News & World Report* devoted a special issue to the topic, "What TV is Doing to America." The journalists' bias reveals itself in several ways. Though the issue contained essays on the role of TV in education and religion, on sports and politics, its lead specialty essay was entitled, "What TV Does to Reading" (39–40), the very preposition, "to," suggesting an aggressive act, very possibly a transgression in the nature of injury. The essay reported on a poll conducted by the magazine that proved quantitatively to the very minute and the percentage point that soon after acquiring a TV receiver, the majority of the public read fewer books, magazines, and newspapers.

The opposition between print culture and television was further reinforced in the special issue, in the interview questions posed to individuals said to be representative citizens in many walks of life. Here is one series of questions: "Do you watch TV a lot? Do you read less than you did before? How about your husband's reading? Many people say they seem to feel that they have wasted an entire evening watching TV—? Do you think TV has made children more . . . knowledgeable? . . . Is it surface knowledge? . . . Do you mean that they don't want to read? . . . Do you mean [your son] can't read a newspaper? What is the answer to this sort of thing, then?" (Sept. 2, 1955, p. 46).

This is not to challenge the specific statistics of a news magazine poll of 1955. It is, however, to notice the ways in which the magazine journalists, by definition a population committed to the culture of print for their livelihoods, their identity, their social status, construct their report and interview in terms of a binary opposition between television and print. They do it so well that it appears to be a natural opposition, as if intrinsic to the sociocultural order. An alternate perspective becomes possible only if the terms themselves are changed, for instance, if one were to pose the challenge to television from another set of concerns—say, by measuring whether time spent daily in spiritual meditation or religious devotion increased or decreased with the acquisition of a TV set, then proceeding to ask about a spouse's devotional and meditational habits, and the children's—or else inquiring into whether the balance had changed with each family member between time spent listening to or playing music *versus* time spent watching television. To suggest other frameworks, other bases of inquiry is to change somewhat the terms of what is considered to be natural. Doing so would expose a seemingly free-form line of inquiry as a highly structured, prosecutorial set of questions containing an underlying argumentative agenda.

Yet the ostensible natural order has long locked the book into combat against television. In 1955, *Coronet* magazine published "I Was Cured of TV: The Story of a Confirmed Addict and His Long, Hard Fight Back to Life," life being defined as reading by the former TV "addict" who returned to vivifying print forms in gradual increments from newspaper headlines to books, putting his personal account forward as a model for others who "might wish to reestablish themselves as human beings" (Batchelor). Again, in 1976, the poet Reed Whittemore argued that television "has reduced reading and literacy [and] increased thumb-sucking and the glassy stare, and it has destroyed what little was left of the American community" (21). If his rhetoric sounds inflammatory, its discursive message has been widely presumed to be true right up to the present, even in graphic arts texts such as the comic strip, "Bloom County," in which the artist–illustrator, Berke Breathed, shows his penguin, Opus, announce from the public library, "No more TV! No boob tube-a-roo!," and continue his rhyme:

> Books! I'll read books! Be they large or quite dinky!
> Straight from the shelves all musty and stinky!
> Faulkner! O'Neill! Twain and Saul Bellow! . . .
> I think I'll curl up with a few of those fellows!

Yet dwarfed by the canyons of shelved volumes in the library stack, Opus realizes his foolishness in thinking the "hour, or three" he has allotted to becoming well-read would suffice—and in the final cartoon panel we see him settled in front of a rerun of "Gilligan's Island" (Breathed, 132). In a similar vein, a reviewer disappointed in Thomas Pynchon's 1990 novel, *Vineland*, complains that the novelist's attack on television misses its mark. "What could be a more rewarding adversary than television," he writes, "providing you actually had something novel to say about the ways in which it alternately stupefies and imbrutes us" (Leithauser, 8).

Given this degree of hostility to television, it is interesting, at least as an aside, to notice the extent to which the manufacturers and marketers of TV receivers deployed imagery emphasizing a congruence with books, and therefore, with the culture of print. To this day videocassettes are marketed in booklike casings, and advertisements and designer showcase rooms have presented the television situated harmoniously with books and with objets d'art, sometimes with a book case built right into the TV cabinet. The books and other objects, as we know, signify high culture and status, which the manufacturers have wished to associate with their TV receivers. But even if the TV screen in an advertisement shows a program in progress, and all books and magazines in the room where the TV set is situated are closed, the images do not place the book and the TV screen in overt opposition or contention. Those with primary allegiance to television have not been moved—and perhaps have not dared—to oppose the world of print. From the early years of the marketing of TV sets, periodicals serving manufacturers emphasized congruence, not rivalry. "A normal furniture grouping serves conversation, reading, or tele-viewing with equal poise" (*House Beautiful*, 91 [Aug., 1949]: 66).

It must be acknowledged also that partisans of print were not the only ones worrying at midcentury about the incursion of television into their realms. In 1950, the president of Twentieth-Century Fox recapitulated the commonly expressed anxiety that "future generations of Americans would be glued to countless television sets," never entering into human fellowship at movies, churches, or civic affairs. And a sports entrepreneur, in Churchillian rhetoric, defended his decision not to televise the Kentucky Derby—"I did not come . . . to preside over the death of the Derby" ("TV the Terrible," 57). To some extent, the promises of televisions' promoters provoked the anxieties in those with business and other interests in areas to which TV made a claim. Promises to make the TV habitat a theatre, movie, sports arena, political town hall, and so forth were quasi-predictions of television becoming a cluster of environments, a multiplex of different worlds that threatened not only to overwhelm the one world of the book, that of "inner space," as a National Reading Week slogan once put it, but to empty out any number of other environments, theatres, stadiums, places of worship.

Yet decisively over the long term the battle has been waged in terms of the culture of print versus that of television. The newspaper critic, Harriet Van Horne, complained in 1951 that "now in the third year of the Television Age, . . . our people are becoming less literate by the minute. . . . Chances are that the grandchild of the Television age won't know how to read this" (quoted in "Dark (Screen) Future"). In 1962, Ashley Montagu referred to "the sterile puerilities which find so congenial a home in the television world." In the name of "good taste," he called for an abatement of its "noisome vulgarity" (132). The vehemence of the rhetoric is clearly indicative of a battle over issues that transcend the question of how the public spends its evenings. And even the rhetoric of Cerf's statement indicates that this author and editor of books was sufficiently anxious about the new medium that he upheld the supremacy of the print form by discrediting the televisual. There is no contest between a good book and a deodorant ad narrated in treacle tones. To enter into the terms of Cerf's statement, or for that matter Montagu's or Van Horne's or the *U.S. News & World Report* journalists', is to know unequivocally where

"Read me."

New Yorker cartoon, 1978. Reprinted by permission of The New Yorker.

correct thought and action lie. Real choice between the book or the television is precluded, as is the possibility of involvement in both. On the one side lie literacy, intellectual curiosity, learning, while on the other are illiteracy, puerility, and ignorance.

At stake in this conflict has been the power and its concomitant prestige on the part of the culture of print. The danger of television, said one novelist–essayist in 1952, derives from its "fantastic *power* . . . it is almost like a giant eye on life itself" (Willingham, 117). One might ask, What eye had traditionally been sanctioned to turn itself on "life itself" if not the novelist's, the poet's and the dramatist's, and the essayist's? At issue was not the prerogative to scrutinize human concerns to the utmost, but rather a fear about the perceived radical shift in the medium of that scrutiny and its practitioners. To read the many print texts is to suspect a certain dread that the scrutinizing "giant eye" would become that of television, leaving the ocular power, the very envisioning power of print culture overshadowed by the newer basis of power and authority in communications. In 1989, the essayist Jonathan Lieberson expressed his fear "that television will swamp what we call literature and that the forms of literary life—represented by book writers, critics, reviewers—will lose the interest and prestige they now have" (19). The image of loss and of being swamped is revealing. To be overcome, even drowned, is the threat seen to imperil the literary establishment.

And if few spokespersons for the print culture are as direct as this, one finds intermittently a similar disclosure of anxiety about power, print, and television. For instance, in a memoir, *The New York Times* columnist, Russell Baker, writes that in the early 1950s he and his wife had no interest in "a world in which people sat silently in the parlor looking at a box that showed little moving pictures" (139). Yet Baker recounts the watershed moment when, in 1960, as a *Times* White House correspondent, he was so busy taking notes on the Nixon–Kennedy debates that he missed learning first-hand that, viewed on the television screen, Kennedy was the clear winner. "That night television replaced newspapers as the most important communications medium in American politics," Baker writes, his insight poignant because, just as he reached the pinnacle of print journalism, television bypassed it in importance. "I covered other debates, but it seemed like just another chore that had to be done because the *Times* was the 'paper of record.' Writing stories, I thought, was like talking to myself" (326). It is the acknowledged threat of obsolescence that most troubles Baker, the sense that he has become a kind of antiquarian, that the "paper of record" is but a daily archive of documents to be buried in a repository and not put into circulation as a dynamic part of contemporary culture. A loss of journalistic relevance and power is central to the anecdote, and rare in the forthrightness of the disclosure.

It must, however, be said that the loss is illusory, because the power and relevance of print never existed at the strength perceived. Most people did not read newspapers, especially in the 1800s. The whole notion that television intervened in the journalist's (and the larger print world's) relevance and timeliness to the vast populace is largely apocryphal, but an effective ideological base for the print world from which to argue against television (see Dicken-Garcia).

But writers' rhetorical strategy divides the public—the print public from the TV public—in terms privileging the one and dismissing the other. Humanistic critics and analysts, speaking from their power bases in the culture of print, have cast their adversarial arguments in high/low culture terms. Their strategy has focused on appeals to class interests, as in a mid-1970s article that begins, "While untold copies of *Ulysses*, their pages yellowed and brittle, sit forgotten in the dark corners of bookshelves throughout the land, *Magnum, P.I.* surges through twenty million homes each week" (C. Anderson, 112). Here the posited opposition between television and the literary text is rhetorically constructed both to privilege and to dismay those with primary commitment to the literary classic. By implication, Shakespeare too lies neglected, along with

Melville, Henry James, Gertrude Stein, and untold others—not only neglected by a mass audience, but abandoned by onetime readers, now turned
traitors or defectors, who are currently content to watch television in mass
solidarity while their books crumble.

These terms of binary opposition carry certain presumptions about the
writer and the public. Typically, the writer and his or her true readers are
construed as the adults, while the TV-watching public are the children—
"spoiled children [who] deny there are problems in life, except more candy
[and] yammer for candy even when it makes their teeth fall out, gives them a
chronic bellyache and covers their faces with pimples" (Willingham, 118).
"Adult responsibility," it need hardly be said, can be found in the culture of
print. One early critic of television worried, he said, not about the "intelligent
person of strong will [who] will become literate, do the necessary studying and
get a good education no matter what happens," but about "most of us" who
"are not strong willed" in an age when "television makes it much easier to be
illiterate" (Utley, 138). Again, there is the opposition between us and them,
though the weak-willed "us," the unstudious and illiterate, have no place
among the real cognoscenti of writer and cohorts. This text even builds in a
cultural litmus test, citing an "old . . . not particularly good movie" on
television whose soundtrack of "Schubert music was lovely" (138). Schubert is
the high-culture test and the signal to the correct reader, who is invited to
identify as a person of knowledge and discriminating tastes. Civilization lies on
the side of Schubert and books, while television attracts the barbarians.

The concern about illiteracy, weak will, childishness is, in this sense, in part
a displaced anxiety about the potential loss of commitment to print and its
texts, a loss both of a medium and of the channels of communication in the
TV age. Predictably, voices from academe have reinforced the bipolar
opposition between literature and television. The most vigorous opposition
has come, not surprisingly, from Departments of English. *Reading in America*
(1989), a collection of essays on the role books have played historically in
American culture, concludes with a lamentation about the status of books in
the television age, focusing especially on "media-induced illiteracy" (285).
Similarly, in *The Death of Literature* (1990), Alvin Kernan writes that "at the
deepest level the worldview of television is fundamentally at odds with the
worldview of literature based on the printed book" (147). Another case in point
is Wayne C. Booth's juxtaposition of televisual and literary texts. Television,
Booth argues, makes immediate and indelible visual impressions but fails to
make any "intellectual demands of the kind expected of even the most watered-
down philosophical or scholarly text, or of the printed fiction that critics take
seriously" (397). In an argument that recapitulates the decades-long normative
stance of the literary intelligentsia, Booth identifies imaginative recreation as

the hallmark of literature, in which "the action takes place in a country somehow in my head, . . . [and is] not confined to a box or screen" (390). Booth excoriates television for forbidding reflection, commodifying the deepest human emotions, stereotyping, presenting a specious gift of programs. His reader, not surprisingly, is required to take the side of literature while acknowledging, anxiously, the power of the encroaching televisual medium.

Fiction, too, has been making a years-long case for itself in opposition to television, embedding within it a subtext arguing for the privileged status of fiction over and against the offerings of television. One good case in point is John Gardner's *October Light* (1977), in which an old woman, Sally Page Abbott, reflects on the difference in kind between dramatic narrative on television and in fiction. Sally's temperamentally rural, elderly brother had complained that television was not true to life, from which premise Gardner launches his, the novelist's, brief for the print text:

> Running through, in her mind, the programs she knew best—*Maude, Mary Tyler Moore,* and *Upstairs, Downstairs*—it struck her that none of those programs even touched real life at all. They were all about interesting characters, stage people, glittering and amusing exactly as characters were glittering and amusing in a Broadway play. That didn't seem to be true, somehow, of characters in novels, even bad novels. If characters in novels were entertaining, it wasn't in quite the same way. They might be a little like characters in movies—a good deal in her [current] paperback reminded her more of movies than life, and perhaps that was why, as she'd known from the beginning, it was trash, really, or at least not the kind of book [her late husband] Horace would read—but there was something, even in a novel like this one, that was more like life than any movie could be. You saw things from inside. You understood exactly why everyone did everything—or imagined you did—so that when something went false it seemed not merely silly but—what? A kind of cheat, a broken confidence. (169)

Here Gardner casts the widest possible fiction net, theoretically encompassing canonical fiction and drugstore pulps, and marshaling the full fictional range as a united front against television (movies too), including the most respected series on commercial and public TV (*Mary Tyler Moore, Upstairs, Downstairs*). His reader is asked to grasp interiority and the analysis of human motives as the fundaments of fiction, and to assent to their importance and agree that they are traits absent in television and film. Popular, trash fiction might—no, would—betray the reader with flaws experienced as falsity, a point that Sally emphasizes at the close of the novel. Even so, it is generically superior to television.

And Gardner resumes this argument, having the woman's brother, the laconic upcountry septuagenarian New Englander named James L. Page, make a polemical statement of a length astonishing to the listeners, his daughter and the hired man—and the reader. Page recounts how for two weeks he watched television, "as fair as any man in a jury box," and saw relentless "filth and corruption," meaning "murders and rapists, drug addicts, long-hairs . . . half-naked women with microphones, stretchin' out their long, limp ahms to you, and puckerin' and smilin' with all their big, glassy teeth . . . quiz shows where people go insane to get some money" (193).

He continues on in this vein, but lest Gardner's reader interpret the speech as the raving of a crank, the novelist hastens to move a mediative figure into prominence; she is the man's daughter, Ginny, who likes television and cannot imagine life without her color set, but, hearing her father's outburst, reflects that "what he said had a fair amount of truth to it." She recalls that once, when their television was returned after two months in the repair shop, "she'd seen it with entirely new eyes. She'd noticed how tiresomely gay things were if they were supposed to be funny, how tiresomely earnest if they were supposed to be mysteries, how program after program had boats or motorcycles in them, as if the same half-wit mind, or eleven-year-old mind, maybe, had written every single story" (197). Gardner uses her as a vehicle to continue to call forth the most routinely used stock lines of TV drama, in order that readers can contrast their verbal impoverishment in the very vitality and freshness of the print text they have in hand ("Walter! Something's happened!" "No use, she's dead!" "Hold your fire!" "Drop it and turn around slow!" [197]). And Gardner concludes the passage of criticism of television with complaints against its commercialism. Ginny can "easily" understand her father's con-demnation of television as evil: the commercials "prostituted children, hard-selling Pop-tarts with a three-year-old's smile. . . . It was all a kind of crime against decency and goodness. . . . You could never again see a white country church . . . without thinking of some mouthwash" (197–98). Sug-gesting the spectrum of television from Public Television through mouthwash commercials, Gardner offers a trio of voices hostile to television and invites readers to recommit themselves to the performance of fictional narrative.

Even the preeminent spokesman on behalf of the television era reinscribes the bipolar opposition between the cultures of print and of television. In the earlier 1960s, Marshall McLuhan, unsurpassed as the most prominent theor-ist of media, cast his ideas and opinions in rhetoric certain to terrify those committed to the culture of print. "Western values built on the written word," he wrote, are in a precarious position because of "a new electronic technology that threatens this ancient technology of literacy built on the phonetic alphabet" (84–85). Unequivocally, he proclaimed the "ecological sweep of the new electric media," especially television, and dismissed those committed to print as old-fashioned conservatives unable to offer even "token resistance."

"The favorite stance of the literary man," chided McLuhan, is "'to view with alarm' or to 'point with pride' while scrupulously ignoring what's going on" (179).

For his part, McLuhan was eager to say exactly what was going on, nothing less than an electronics–telecommunications revolution certain to doom the culture of print as it was understood by anyone in the 1960s. McLuhan foreclosed any possibility that two cultures, print and telecommunications, could co-exist. "A new medium is never an addition to an old one," he wrote, "nor does it leave the old one in peace; it never ceases to oppress the older media" (158). He chided those in the print culture who lack "cool visual detachment [and feel] panic about the threat of the newer media." But this professor of *literature*, by training and profession a man of letters, must surely have understood the impact of his rhetoric on his professional colleagues and on their cohorts in journalism and other print forms when he reminded them that their predecessors, the medieval scholastics, were driven to extinction by the printed book, and when he then proceeded to declare the new electronic media, especially television, to constitute a sweeping epistemological revolution "vaster in scope than that of Gutenberg" (162). In his rhetoric, McLuhan both named and also nurtured the "current anxieties," even the "moral panic . . . of civilized man concerning the written word" (84). Television is so difficult a subject for literary people, he argued, that it must be "approached obliquely" (150). He dismissed the obsolescent print culture as one of "earnest moralizers," declaring epigrammatically, "A moral point of view too often serves as a substitute for understanding in technological matters" (216). All these pronouncements thread their way through McLuhan's text, eroding the very terms on which print culture presumed to stand and limning the obituary for its practitioners.

Close readers of *Understanding Media* must nevertheless detect McLuhan's own anxieties, expressed, for instance, in imagery of nuclear devastation. "Education," he wrote, "is ideally civil defense against media fall-out" (175). Evidently worried about the monopolization of television by commercial interests, he spoke of the new media as constituting "collective surgery carried out on the social body with complete disregard for antiseptics," thus raising the possibility of "infecting the whole system" (70). Perhaps it was this anxiety about electronic—especially televisual—infection–radiation that led him in places to sound authoritarian. Like a behaviorist, he hoped that in the TV era "whole cultures could now be programmed to keep their emotional climate stable" and spoke of the intellectual in the electronic age as a figure of "social command" holding the "whip hand" in society (41). McLuhan continues to attract proponents and to provoke skepticism and condemnation. He interests us here because he exacerbated the bipolar oppositional terms between print and television. Rare among commentators in his implicit championing of the telecommunications era, McLuhan nonetheless continued the argumentive structure of the kind Bennett Cerf had employed years earlier, and that virtually every text reiterated when confronting the relation between television and print forms.

And although McLuhan's bipolar opposition favored television, the same structure of argument has come in for reprise much more recently, this in a text that draws the opposite conclusion from McLuhan's but once again reinscribes the same set of terms. Neil Postman's *Amusing Ourselves to Death* (1985) warrants attention here precisely because of that reinscription. Postman, a specialist in communications, insists that television "keeps us in constant communion with the world," but his essential argument is that "television . . . has made entertainment itself the natural format for the representation of all experience" (87). Rhetorically, Postman fortifies his position by referring to the telecommunications era in the infantilizing term, a "Peek-a-Boo World" (64–80). His polemic castigates television for its trivialization of all human experience. Against this contemporary degradation of American culture, he posits its bipolar opposite, an American past that he casts in a utopian aura of reason, rationality, and verbal complexity. This is "typographic America" of "the typographic [American] mind." Its traditions reached, in Postman's construction of history, from Plato and Moses through the *Mayflower* Pilgrims and thereafter to Benjamin Franklin and the eighteenth-century Enlightenment of the Founding Fathers, hence it proceeded into the nineteenth century of the seven-hour political debates between Abraham Lincoln and Stephen Douglas (30–63). The text enshrines this era as one of literacy, print, and rationalism, all three cohering in an ideal triad. Literary, linguistically and discursively complex, rich in content, this entire period is construed as one of typographic values enacted in every quarter of American society, including politics, religion, jurisprudence, even advertising, which, "as late as 1890, [was] regarded as an essentially serious and rational enterprise whose purpose was to convey information" (59). Once again, one finds the idealization of a mythic world of print, but one that was in reality inhabited by a minute fraction of the public.

The point is not to disprove Postman's premise or to decide who is right, he or McLuhan. And whether one finds the pretelecommunications American past of *Amusing Ourselves to Death* to be accurately objective or idealized, one must recognize its place in the continuing pattern of discourse that is locked in the bipolar oppositional mode. Postman's text is the obverse of McLuhan's; they are contraries. Yet their texts are replays of each other, entirely congruent in their bipolar oppositional structure. Their emphases differ; their propositions are identical.

What about those in the culture of print who cross the barricade, moving into the land of television? The crossover is complicated, given the way in which the world of books claims the values of ethics, the arts, civilization. How then can a citizen of that country explain a sometime foray into television? The answer becomes, through sickness and debilitation. The late poet L. E. Sissman, who earned his living in advertising, said he watched television "in massive, concentrated dose" only when too weakened by illness (and the medically induced illness of chemotherapy) to lift books. Even so, he wrote that late in life, attempting to watch a new show with his wife, he grew nauseous and became a victim of "acute television poisoning," staggering off to bed where he collapsed in the position his print culture cohorts would approve: he "fell down under a good book" [25, 27]).

Similarly, when statistics in 1973 showed that Americans watched an average of four hours of television daily, an editorial in the New Republic protested, "Four hours a day! The average American lives some of his life as if he were indeed in the hospital, and that kind of living surely is a real effect of TV" ("Sedulus"). More recently still, a critic reporting on daytime television for The New York Review of Books, began his essay with a comparable apologia of illness, explaining that he already knew something about daytime television only because he had once been hospitalized ("seriously ill . . . fitted out with catheters and intravenous tubing") and, therefore, so debilitated that it was not possible to read. Though he had been "immoderately" drawn to television while hospitalized, the "obsession" afterward subsided. Now, solely in order that he might consider writing a critical essay on the subject of daytime television and its epistemological shaping of public consciousness has he decided "to spend a day in bed watching all the shows I used to watch" (Lieberson, 15). Readers of his essay will suspect from its very detail and from the comprehensive, multichannel coverage of daytime TV that the alleged "day in bed" is a narrative device, rather like the days in the Book of Genesis.

Yet the "day in bed" is symbolically significant. The only justifiable way for the print culture person to watch daytime television is to take to his bed, that is, to mimic the posture of illness, self-consciously pantomiming sickness, that is, disability, so that his readers understand he has not really actively chosen to spend time watching television. The print culture person can only legitimate TV watching by bracketing it outside of normal life. Because the fellowship of the print culture "forbids" daytime TV watching, it can only be justified as a temporary aberration—and not a mental–intellectual one but a physical one over which the mind has no control. Physiological morbidity becomes the permission slip for TV watching.

Essentially, then, the two cultures of television and of print have been cast in the adversarial principles of puerility versus maturity, low culture versus

high, entertainment versus intellectual engagement, frivolity versus serious-
ness, contamination versus purity, robotry versus critical imagination, sickness versus health. These are the battle lines that have continued, uncon-
tested, for nearly forty years. The way the argument has been cast suggests that
resolution of the conflict is possible solely in one's maturation. The child held
captive by the powerful popular medium will ultimately renounce television
and embrace the intellectual–imaginative complexity of the printed text. As
one analyst observes, "We expect children to like television precisely because
they are easily amused and do not know any better, but we also expect them to
grow out of it" (Attallah, 225). The enlightened adult can be expected to cross
the oppositional boundaries, while the boundaries themselves remain intact.

So the seismic fault lines have continued for over four decades, grinding
against each other like argumentative tectonic plates. Simultaneously, however, in very recent years there are clear signs of a change of position, of the
argumentative barricade between the book and the television falling in some
quarters—for instance, as we see an occasional cartoon in which an individual
sits before the glowing television screen with an opened book in his lap, or as
characters in novels of the 1980s are portrayed *both* as readers *and* as television
watchers, for instance appreciating James Baldwin's *Nobody Knows My Name*
and Gabriel Garcia Marquez's Latin American classic, *One Hundred Years of
Solitude* as well as "The People's Court" and Johnny Carson. Or as *The New
York Times* runs a feature story on the extent to which television (and movie)
references are proliferating in novels by younger writers. Or as the public
institution most heavily committed to print, the National Endowment for the
Humanities, in 1988 commends "using the image to increase appreciation for
the word" and quotes approvingly Daniel Boorstin's remark that "people watch
television *and* they read," (McMillan, 81, 138, 310, 65; *The New York Times*,
May 31, 1990: C15, 19; *Humanities in America*, 19, 20).

This kind of configuration is radically different from the one on thumb-
sucking and glassy stares, brutishness and so forth, and it shows that the older,
heretofore impermeable boundary is becoming porous. This, then, is precisely
the point at which it becomes possible to analyze in historical terms a conflict
articulated for decades by the print world as the representation of sociocultural
truth. Only now does it become possible, that is, to suggest that there is a
politics embedded in this battle, that there have been personal and political
interests at work in a power struggle waged for the control of meaning and
medium. The decades-long intransigent and vocal opposition to television by
those with primary commitment to print texts has been an American
twentieth-century battle of the books—or rather, a battle by the books carried
out in the very bipolar oppositional terms set forth by Bennett Cerf in 1948.
Only very recently has the bipolar opposition diminished as a new, second

generation print-culture population experiences television, not as a territorial or intellectual threat, but as "part of the very air [they] breathe" (James, C19). The either/or model begins to yield to the larger, inclusive and conjunctive structure of both/and.

Like so many conflicts, this battle of the books is evidently being abandoned before it is resolved—and by the very TV-generation erstwhile "children" who in the 1970s–1980s are refusing the old terms of argument. Not that the writers refuse maturity; they are not a generation of video Peter Pans. Rather, they reject the division between the worlds of television and print text. The author and wit, Roy Blount, Jr., for instance, describes the act of reading serious contemporary fiction in terms of teleconsciousness, his dust-jacket blurb praising Lee Smith's novel, *Black Mountain Breakdown*, in this way: "The closest to reading this would be reading *Madame Bovary* while listening to Loretta Lynn and watching 'Guiding Light.'" Here the multivalent, media-age mind of the reader is presupposed, not only invited into a fictional world combining country music, TV soap opera, and the canonical French classic novel, but one simultaneously audible, visual, and linear. The contemporary reader is urged to seek out Smith's novel precisely because it enacts cultural variety and operates teleconsciously.

And the writers commit themselves to this new inclusiveness. The narrator of a William Warner story remarks, without embarrassment, that "some writers *first* trained by reading Dickens or Fitzgerald . . . others by watching TV" (197). Warner, among others, rejects the position that television is alien, even inimical to the literary imagination. He and his cohorts simply refuse those terms of engagement and the hierarchy implicit in the terms. Lawrence Wright's memoir, *In the New World: Growing Up with America 1960–1984* (1988), for instance, intermixes lifelong watching of sitcoms, TV news, congressional hearings with avid reading of Joyce, Bellow, Percy, Richard Wright. Altogether, these experiential references constitute the autobiographical natural world. Including television and literary texts in this way, Wright shows the newer fiction writers' preparation to exploit the cognitive traits of the television in fiction, for instance, adopting the televisual flow identified by Raymond Williams as a form better suited to the contemporary cognitive reality than the erstwhile beginning–middle–end structure (see Tichi).

It is helpful to see in some detail how fiction itself is now arguing that literary and TV texts are conjunctive in the contemporary American consciousness. One recent novel, Jill McCorkle's *The Cheer Leader* (1984), presents at length the holistic relation of television and literary texts. It mixes the two freely, even promiscuously, to represent the contemporary consciousness of the writer. In this coming-of-age novel, Joslyn (Jo) Spencer, a high school girl, bursts the cocoon of slumber parties and small-town life during a summer involvement with an older boy before she goes off to college. McCorkle's protagonist, an avid reader of classic literary texts, is perfectly positioned to expose the relation of the literary mind to the realm of commercial television. At home in the world of books, Jo becomes a significant TV test case precisely because she is not an unschooled, inarticulate character susceptible to uncritical acceptance of the medium.

In *The Cheer Leader*, literary and TV culture cohabit comfortably. McCorkle's protagonist is a young woman of words, an aspirant poet who reads Proust, Maupassant, Dickinson. At the same time, her lifelong frame of reference features television. A friend wears a bikini and "I Dream of Jeannie" pants, while a boyfriend looks "as All American as Wally Cleaver on Leave it to Beaver." At first sight, Jo's boyfriend-to-be does not have a 'Then Came Bronson' ruggedness. "Clearly," she says in an attempt to define herself, "I am not an 'I Love Lucy' nor am I 'That Girl'" (36, 46, 59, 60, 84, 264). These references come interspersed with serious statements on such canonized writers as Chaucer or Plath or Sherwood Anderson or Proust. All comprise one unitary world; there is no disjunction between canonical literary and TV texts.

Moreover, television is embedded in childhood's best memories (and we doubtless can expect more of this in quasi-autobiographical fiction of the foreseeable future). Jo recalls childhood wrapped in "flannel jammies," when she would "sit on the floor and let the dog chew on the fuzzy slippers that came last Christmas," and rest her head "on Daddy's knee and watch the gray TV light flicker on the knotty pine paneling, watch every fear of the day dissolve into the gray, into the hum and lullaby" (170). The maternal side of this experience is here too, the late afternoons, "the TV on, black and white, her mother ironing . . . [while watching] 'As the World Turns' while she and [her brother] Bobby eat vanilla wafers . . . until . . . her mother is cooking dinner and Andy Griffith is on the TV" (182).

Projecting her future, Jo never repudiates television, never hints at its supersession. Chaucer and "I Love Lucy" exist simultaneously. They are not even juxtaposed, not set in self-conscious relation to each other. True, this kind of consciousness flattens history into one present state. But there is no irony in McCorkle's positioning, no polemic statement about high culture versus the popular. Jo is not slumming in TV reruns.

And she is not, as a poet, devotee of words, planning to put aside, much less to reject, television in adulthood. At the end of the novel, imagining marriage and a young son, to be named Anaximander, she envisions rainy days when "I will roll back my nice oriental rugs and little Anaximander and I will roller skate while we watch old reruns on TV" (265). This projection of the future will be a permutation of her own childhood experience. The fantasy is based on the presumptive pleasure of television. In the consciousness of this poet–

protagonist a Greek philosopher namesake and TV reruns coexist in perfect accord. Literacy, literature, intellectual life, McCorkle argues, include popular television. *The Cheer Leader* is significant here precisely because it shows how television has been naturalized and how it has entered the contemporary American literary consciousness and thus the culture of print. Texts like McCorkle's (and Wright's and Warner's) argue that a new era may be upon us, one in which the very idea of the two cultures will recede and be understood in retrospect as a temporary period of technological accommodation and its abrasions, though it is entirely possible that the print culture will divide into factions, some recycling the established arguments while others integrate TV forms and motifs into their texts.

Nor is this a phenomenon confined to those print culture writers affiliated with the academy. Though McCorkle and others, like Oscar Hijuelos, whose *Mambo Kings Play Songs of Love* (1989), based on a connection with Desi Arnez on "I Love Lucy," may preside over writing workshops or enjoy having their books taught in college and university classes, any reader of mass-market mysteries can report that television has been naturalized in those texts too, for instance in Stephen King's books or the mysteries of Sue Grafton, Sarah Peretsky, or Sharyn McCrumb whose mid-South detective does an Andy Griffith act when useful to him, or Ann Rivers Siddons, one of whose plots turns on a TV movie.

In one sense, a question might be, Why has it taken so long? Why only in the 1980s, a full quarter century into the TV age, has a segment of the print world seen its own interests served by amalgamating television and print. A younger writer, David Foster Wallace, answers that question in intergenerational terms. Older writers whose consciousness was formed in the pre-TV era, he says, dismiss "strategic reference to 'popular culture' . . . as at best frivolous tics and at worst dangerous vapidities." He cites his own onetime professor of fiction who proclaimed that, to attain classical timelessness, "serious" or "literary" writing must omit the popular ephemeral. Wallace's classmates were quick to point out that the professor's own fiction abounded with electric lighting and automobiles. But when pressed, the instructor said he drew the line at "'mass-commercial-media.'" Buicks and Oldsmobiles, yes; the Ricardos and Bradies, no. In the class, Wallace reports, "transgenerational discourse" broke down at that point. But implicitly he also tells us why the print culture has been relatively slow to exploit television. The authorities, whether professors, editors, supervisors in various positions—in short, the gatekeepers—have kept vigilant watch, discouraging and forbidding the integration of the two media worlds until overpowered by the young, now adults, for whom separation of the two would be deeply unnatural.

The Child—A Television Allegory

Tomorrow's children, through the great new medium of Television, will be enrolled in a world university before they leave their cradles. . . . Think what this means. How splendidly equipped they will be while young . . . to carry the torch of civilization forward into undreamed-of fields.
Allen B. DuMont Laboratories Advertisement, 1945

Most parents are aware of the potential dangers of letting young children watch violent TV shows, but many don't realize that violence isn't the only problem. . . . Too often both children and adults turn off their brains when they turn on the set.
Melitta Cutright, *The National PTA Talks to Parents*, 1989

Instead of being overwhelmed by Cyrus, Theseus, Moses or Romulus, [children] unconsciously act out the roles of the . . . TV personalities around them. One can only pity young people . . . who are artificially restrained from the enthusiasm for great virtue.
Allan Bloom, *The Closing of the American Mind*, 1987

No aspect of the acculturation of television has remained at once so vigorous and so static as the debate over the relation of children to television. The television child is precocious and sophisticated, according to the TV industry, and certain educators and child development experts cautiously agree. Yet that same child is otherwise profiled as a hostage to TV violence, as a child developmentally hastened into pseudomaturity, as a television "wolf child" suckled by television and cut off from the "real" world.

The debate goes on as ritual, not really as argument. The terms, set in the 1950s, reappear as a litany repeated over decades. A massive social science literature has reported the inconclusive results of multimillion-dollar research projects on the effects of television on children, and still journalism on children and television is virtually identical from one year to the next, the perennial recycling of the same configuration of issues itself arguing a continuing, unresolved social anxiety. Television is recurrently constructed both as an intractable problem and an educational opportunity for children. If we no longer read ophthamalogists' warnings that television might harm children's eyesight, a concern of the early TV years of the 1950s, a panoply of other dire warnings remain in place from those years, that television lures children away from family concerns, from school work, from "wholesome" recreation, that it glorifies violence and, ruining school grades, leads to juvenile delinquency. On the one hand, the TV-age child is construed as a toddler with baccalaureate degree, on the other a literal enfant terrible or stupefied isolate identifying with TV role models rather than with history's male emperors, sages, and founders of nations.

To some extent, the child of the TV era is a surrogate for, and an extension of, the same conflicts that have beset adults struggling to accommodate themselves to the new technology. The conflict between the culture of print and the culture of television, for instance, involves the TV child. Benjamin Spock's advice manual on parenting cautions that "nowadays television often takes the place of reading," and Spock warns parents to limit children's TV watching so they can "have time to develop their own imagination" and express it in their creative play (283). A spokesperson for the PTA similarly warns that "television is the main competition for reading, so any campaign to get your child to read more will undoubtedly include turning off the TV for much of the evening" (Cutright, 110). She adds, "Look for TV shows based on books" (111). Here the television child is positioned like the adult, caught between the screen and the books. The conflict, displaced onto the child, is replayed in the same terms.

To some extent, in addition, cultural preoccupations publicized in print and broadcast media inevitably focus on the TV child—but actually reflect the social issues of the moment. For instance, the mid-1970s public revulsion against the Vietnam War arguably prompted these titles—"Saturday Morning Massacre," "Combat Diary from a Battle Zone"—attacking the violence in Saturday morning cartoon programs but in so doing revealing misgivings about the involvement of the United States in Vietnam (*America*, June, 1974; *Progressive*, Sept., 1974). And the mid-1980s preoccupation with health and physical fitness emerges in such titles as "TV Makes Fat Kids," "Widebodies" (on TV watching and children's obesity), and "Like Parent, Like Child: Turn off the Tube, Turn on the Exercise"—all appearing in the years of particular popular emphasis on personal wellness and on exercise as a means to achieve it

Saturday Evening Post illustration, 1968. Reprinted by permission ©
Curtis Publishing Co.

(*Prevention*, Sept., 1985; *Health*, Oct., 1985; *American Health*, July–Aug.,
1988).

Still, the constant journalistic incantations on the TV child from the 1950s
to this day indicate that the TV child is a character in an ongoing American
allegory. Like all allegories, the characters personify abstract qualities, and the
action and settings represent the relationships among these abstractions. The
children are the Innocents on a pilgrimage, the parents the Guardians who
must at times seek guidance from educators and social scientists, the Experts,
who can advise them on how to mediate between the Innocents and television.
Television in this allegory is a metamorphic character, sometimes the Illuminator, at other times the Corruptor. The parents bear the burden of knowing

which TV identity is uppermost at any given moment, since the Illuminator
and the Corruptor can metamorphose into each other practically at the speed
of light with the change of an image.

Like all allegories, this one exhibits objects and persons who are equated
with meanings that lie outside the narrative. The Innocents are pilgrims whose
ultimate success, both spiritual and material, reflects the quality of their
parental care. In turn, the parents' worth is measured according to how well
they, as Guardians, fend off the corruption of television and yet facilitate its
illumination. The stakes are very high, for if the Guardians fail, their pilgrim–
Innocents will be destroyed. The future of the nation is ultimately at issue.

The allegory takes place in a realm—The Home—presumed in the pre-TV
era to be a peaceable kingdom of harmony between parents and children. With
the installation of television, however, the schizoid Illuminator–Corruptor,
The Home becomes a problem zone. Whereas the Innocents and Guardians
lived in harmonious reciprocity of love and nurture before the TV era, the
introduction of the receiver into The Home essentially forces the erstwhile
parent–child relationship into triangulation. Television becomes a third
vector, commanding the children's attention directly, usurping the prerogatives of the Guardians, who must now struggle to retain or recapture their
authority.

And the moral fable enacting this very point is available for children as a
cautionary tale. Betsy Byars' *The TV Kid* (1976), features eleven-year-old
Lennie, who lives with his mother in a motel she owns and runs. Lennie is
shown to be lonely, bored, failing science—and immersed in television,
especially game shows in which he is the fascinated consumer. As the novel
opens, he is hosing down a walk, reminded by his mother that his homework
(again, homework, one province of the Guardians in the TV era) must be
done:

> "Aren't you through yet?" Lennie's mother called. "You've got to do your
> homework, remember?" . . .
> Lennie walked on to the office. As he went inside, he paused in front of the
> TV.
> A game show was on, and there were five new cars lined up on a revolving
> stage. The winning contestant got to pick one of the cars, and if it started, he got
> to keep it. Only one of the cars was wired to start.
> "It's the Grand Am," Lennie said instantly. He felt he had a special instinct
> for picking the right box or door or car on shows like this. "I *know* it's the Grand
> Am."
> "Lennie, are you watching television?" his mother called from the utility
> room.
> "I'm looking for a pencil," he called back.

"Well, there are plenty of pencils on the desk."

"Where? Oh, yeah, I see one now."

Lennie was hoping to stall until he could see if it really was the Grand Am as he suspected.

The contestant wanted to try for the Catalina. "No, the Grand Am, the Grand Am!" Lennie murmured beneath his breath. He found the stub of a pencil on the desk. . . .

"Lennie, I meant what I said about no television," his mom called.

"I know you did."

"No television at all until those grades pick up."

"I know."

A commercial came on. "Doc-tor Pep-per, so mis-under-stooooood."

Here we participate in Lennie's transit to and from the seductive television world of the game show and of his Guardian–mother's demands, back and forth from the directives of the Guardian to the corruptive world of television, which, until the end of the novel has the higher priority for the boy. Ultimately, a crisis—a rattlesnake bite that threatens Lennie's life—brings renewed, intimate contact with his mother and with the policeman who rescues him. Watching television from his hospital bed, Lennie has a TV epiphany, realizing "the people who made television commercials didn't know anything about life. . . . It seemed to him suddenly that every TV person he had ever seen wasn't real. . . . Lennie thought, his own family—just him and his mom—was a hundred times realer than the Bradys or the Waltons or the Cleavers or any other TV family you could name" (108–9). This is the revelation that rescues Lennie from television–the–Corruptor and brings him back into the world of The Home and Guardian. Within moments, he switches off "Let's Make a Deal" and begins work on his science report. He has come home.

Yet the allegory continuously reinscribed in public discourse does not conclude so neatly, snake bite aside. The Guardians get no serpents ex machina in child-care manuals, education texts, or journalism. Instead, to win and protect the Innocents, they must position themselves as mediators between the television and the children. They must assume an intense vigilance, promoting illumination, forefending corruption. They cannot rest. A sampling of essays in middle-class periodicals suggests these tensions of the TV era, as the corruptive side of television is configured as a "bugaboo," a "menace," a "pest" like vermin, needing extermination, a "monster," a "time-bomb," a narcotic subject to "overdose." This list shows television as an infestation, as an explosive, addictive drug, monster—all images that emphasize danger from without, invasion and attack that leave the Guardians and the Innocents vulnerable to injury, or worse (*Today's Health*, June, 1950; *PTA*

Sony Corporation advertisement, 1970s.

Magazine, May, 1971; *Parents Magazine,* June, 1977; *Better Homes and Gardens,* Sept., 1988).

But the Guardians have the help of the Experts, as the same middle-class periodical press indicates in such titles as, "Is Television Good or Bad for My Children?," "Should You Tear 'em Away from TV?," "What Shall We Do About TV?," "Is TV Running Your Household?," "Should Children's TV Habits Be Controlled?," "TV and Family Life? Do They Mix?" (1950, 1955, 1968, 1987). Characteristically, these articles encourage family viewing for family cohesiveness and otherwise urge parental monitoring and curtailing of children's viewing, which is to say that they advise vigilant and active guardianship.

Essentially, then, television is configured in the allegory in quasi-religious terms in a polar opposition between evil and good. Television, as a meta-morphic character, is both satanic and, as a messenger of divine offices, is angelic, even in a secular age. But the Guardians must stand watch to discern which incarnation of television is present to the children at any given moment, and act as censor of the corruptive temptation or as facilitator of the light. Thus the "good" programs and the "bad" ones, the former clustered and approved by the Experts' consensus: "Make good use of TV, encouraging your child to watch such shows as 'Reading Rainbow,' '3-2-1 Contact' and 'Square One'" (Cutright, 160). The "good" programs form a 1950s–1990s inventory from "Kukla, Fran, and Ollie," through "Captain Kangaroo," "Mister Rogers's Neighborhood," "Sesame Street," "The Electric Company." Though each has drawn some criticism (for instance "Sesame Street" and "The Electric Company" cited for reinforcing a shortened attention span or being in format too much like the commercials), in the main these are the "good" programs (see Lesser, 174–201). Action for Children's Television (ACT) is prominent as a paragon of children's programming (Lesser). Their programs, and the others named, are seen to promote healthy curiosity, sharing, ethnic and racial pluralism, ingenuity, egalitarianism, literacy, imagination, and creativity.

The "good" programming is also cosmopolitan, broadening the known world of the young Innocents, preparing them for an eventual knowledgeable adulthood. "Children who grew up before [TV] was available labored under an ignorance which now seems incredible," says one child-rearing expert, who goes on to describe this pre-TV child as profoundly ignorant of other racial groups, other speech patterns, other geographical areas, other workplaces. "Many children simply did not know there *were* people who talked differently from their neighbors; families who lived differently; people who worked at jobs not locally available" (Leach, 677). The pre-TV children in this sketch are so provincial and insular that anthropologists might have wished to study them as locally pure culture.

And the TV industry has exploited this very point. The 1970 advertisement for *May's Child*, the term borrowed from the classic verse about horological attributes of children, features a boy of about ten in a bathing suit with snorkeling gear (mask and fins), slung with canteen, camera, rope and climbing pick, his right arm encircling the neck of a cello. He is clearly the contemporary Renaissance boy, an undersea and mountaintop explorer, camper, photographer and arts activist. He swims, climbs, shoots (with shutter, not trigger), and bows his viol. And surrounding him in the text, sponsored by the television industry trade organization, the Television Infor-mation Office, are the listings of May, 1970, programs to make him a participant in all the roles signified in his assemblage of equipment. Environ-

May's child.

What is television doing to our children?

here are some who say television ives children strange ideas. Right. nd here are some ideas your children ay get from television in May.

They might be intrigued by ocean-graphy or marine biology. Maybe be-ig a potter, or a photographer, or a culptor will catch their eye. Perhaps ley'll be led toward conducting, or riting.

Of course, May brings bitter as ell as sweet. Your children will see hat we have done to our world, and won't all be pleasant. They'll find ut what ecology is, and what it can lean to the rest of their lives. Tele-ision feels bound to report to your hildren the bad as well as the good. iecause they—and you—have the right o know and the need to know.

So for May, then, some programs o open young minds.

Teacher, Teacher. Hallmark Hall of Fame—Drama about mentally-retarded youth, vith David McCallum, Ossie Davis, George Grizzard. Saturday, May 2 7:30-9 pm).

Discovery. The Young Oceanographers; Thundering Waters of Niagara; Lights, Cameras and Underwater Adventure; Strangest Mammals of All; The Great Salt lake. Sundays, May 3, 10, 17, 24, 31 (11:30-12 Noon).

In Which We Live. New weekly series will explore environmental situations and pre-sent concern of Americans with ecological problems. Sundays, May 3, 10, 17, 24, 31 (5:30-6 pm).

The Unseen World. The infinitesimal, the very distant, the very quick and the very slow are highlighted in Isaac Asimov's script. Sunday, May 3 (8-9 pm).

Now. Black Mayor in Dixie; Eye of the Storm; Missing in Randolph; Women's Liberation. Mondays, May 4, 11, 18, 25 (10:30-11 pm).

Once Before I Die. GE Monogram—Report of seven amateur mountaineers climbing Himalayas. Tuesday, May 5 (7:30-8:30 pm).

Cartier-Bresson's California. Impressions of section of American life by French photographer. Tuesday, May 5 (10-11 pm).

With These Hands: Rebirth of American Craftsman. Ideals, viewpoints, creative insights of eight Americans who have turned from assembly line methods. Friday, May 8 (9-10 pm).

Those Incredible Diving Machines. Under-sea World of Jacques Cousteau—Docu-mentary on man's drive to master ocean depths. Saturday, May 16 (7:30-8:30 pm).

The National Environment Test. Americans may test their knowledge of ecology, with Harry Reasoner, Mike Wallace. Tuesday, May 19 (10-11 pm).

Mission Possible: They Care for a Nation. Examines environmental deterioration from a national viewpoint. Wednesday, May 20 (10-11 pm).

The Great Barrier Reef. Exploration of Australia's spectacular shield against the ocean. Friday, May 22 (7:30-8:30 pm).

Wilderness Road. American Rainbow—10-year-old boy studies Arizona's wildlife with his uncle. Saturday, May 23 (12 Noon-1 pm).

The Anatomy of a Symphony. New York Philharmonic Young People's Concerts—Leonard Bernstein conducts and narrates in a performance of Respighi's "Pines of Rome." Sunday, May 24 (4:30-5:30 pm).

60 Minutes. Tricia Nixon conducts tour of White House. Tuesday, May 26 (10-11 pm).

The Shining Mountains. Project 20—The people and legends of America's Rockies. Wednesday, May 27 (10-11 pm).

Regularly Scheduled Programs

Monday through Friday: Sunrise Semester/The Today Show/Captain Kangaroo/Sesame Street/Misterogers' Neighborhood

Tuesday: First Tuesday (appearing first Tuesday each month)/ CBS News Hour

Saturday: The Banana Splits Adventure Hour/Jambo/ABC's Wide World of Sports

Sunday: Lamp Unto My Feet/Look Up And Live/Camera Three/Bullwinkle/Discovery/Guideline/Face the Nation/Meet the Press/Directions/Issues and Answers/G-E College Bowl/Wild Kingdom/The Wonderful World of Disney

Note: This is, necessarily, a partial listing. Time (NYT), titles and casts of these national programs are subject to change. Please consult your station listings; check also for noteworthy local programs.

Television Information Office
745 Fifth Avenue
New York 10022

Television Information Office advertisement, 1970.

mental exploration figures prominently in ten programs ranging from ocean-ography to Himalayan climbing to an "exploration" of Arizona wild life and Australia's Great Barrier Reef. Otherwise, there is a program on the rebirth of American crafts and on the "anatomy of a symphony." Descriptions of these programs literally surround the boy, making him an activist in the sports of discovery, and the arts. "What is television doing to our children?" asks the tantalizing headline rhetorically, answering its own question by promising to "intrigue" the (presumably male) child viewers: "they might be intrigued by oceanography or marine biology" or "maybe being a potter, or a photographer, or a sculptor will catch their eye" or "perhaps they'll be led toward conducting, or writing." The purveyed notion is that the televisually "intrigued" child will be motivated to go forth, an activist, into the arts and the global environment. May's child is a burgeoning cosmopolite.

And the novel, too, can join the advertisement and the child-rearing manual in discourse arguing on behalf of "good" programming. For instance, Norma Rosen's *Touching Evil* (1969) advocates public programming as the invaluable history lesson for the moral development of the young. Set in New York City, *Touching Evil* addresses race and class division in the United States, and its plot includes rape, robbery, and marital betrayal. But the crux of the modern conscience in the TV age lies in the televised trial of the Nazi war criminal, Adolph Eichmann.

Daily at five P.M., Hattie, a pregnant young wife—"a child with a child inside"—sits down in her cramped Manhattan apartment to watch a trial episode. The narrator of *Touching Evil*, a woman in midlife, tells us she herself "lived through the whole business the first time around." Now, protective toward the younger child–woman, she watches the televised ses-sions with her. Hattie, she sees, is "watching the Eichmann trial on TV [like] a frightened child who's turned on the horror movie and doesn't know how to turn it off."

In theme, the novel asks how life can be nourished in the face of evil, and the telecast trial is a window—a source of Illumination—on evil unsurpassed in Western history. The programming becomes a conduit of conscience for the later twentieth century. Television is a school, and Rosen's narrator puts it in these very terms: "We sit on hard kitchen chairs drawn up before the TV, watching, as if putting ourselves through school." The TV trial, showing film clips of Nazi atrocities, brings the horrors with the regularity of weather—"like rain, like cold. The machine guns punching bullet holes, the clubs beating against bone. The starving are starving, the screaming are screaming" (28, 43, 51–52, 68–69, 75, 82, 86–87, 108, 116, 154–55, 209, 217–18, 258).

The novel affirms the morally educative use of television for the young. Hattie will learn about the Holocaust, see the images of the concentration

camp, hear the eyewitnesses. Hattie's brother-in-law, irked by her preoccupation with the televised trial, reminds her that human pain "was always there,
Hattie, even before TV." He is right but misses the point, which is that
television–the–Illuminator is the contemporary means of initiation into
knowledge. The televised images of camp witnesses do not fade but "hang in
the air, continued" in the viewer's "own self." Hattie "drinks in the words,
sucks up the images. This is watching TV as TV means to be watched, as
children watch."

As for the facts of the trial, the narrator gets them daily from the newspaper.
But she knows *The New York Times* runs a poor second in impact to the
television environment: "It doesn't do any good when I tell [Hattie] that I've
read my paper. She has watched living clips on TV. She is in the news and the
news is in her." The TV environment is interpenetrant here, viewer and
television becoming one. And Rosen approves in educational terms. Such
programming is good, she tells us, because it can school the sensitive, those
who bring new life into the world, who ask the ethical questions, who become
the conscience of the modern world. (As a point of contrast, Hattie's husband,
a photographer and thus recorder of visual images, is preoccupied with
composition but uninterested in moral issues, while her brother-in-law, a
salesman of TV sets, never watches the trial.) Rosen's novel argues that
television is at its best as a school, that is, instructive, bearing knowledge,
engendering ethics.

But while television is such a powerful force for good, the other side of
television, of corruption and debilitation, always lies in wait, as the Experts
warn. "The amount of time most youngsters spend in front of TV is
appalling . . . research has found that children who watch a lot of TV get
lower grades, put less effort into schoolwork, have poorer reading skills, play
less well with other children and have fewer hobbies and outside activities than
do children who watch an hour or less of TV per day" (Cutright, 115). Even
"May's Child," the precocious cellist and skin diver, can be seen to lead an
unnaturally accelerated life, if his parents mistake the so-called educational
programs for Illumination, when in fact they represent the corruptive side of
television. "May's Child" is "the hurried child," says one expert, rushed into
pseudomaturity in part by the most sophisticated of TV programming. He is
the child who can (at least after May 31) "talk about nuclear fission, tube
worms at 20,000 fathoms, and space-shuttles." He is also seemingly "knowledgeable about sex, violence, and crime." All this, however, is "largely
verbal . . . a pseudosophistication which is the effect of television hurrying
children" and that "encourages parents and adults to hurry them more"
(Elkind, 78–79).

Most "bad" programs, not surprisingly, are linked to violence, to the idea sustained in the social sciences that children will model their own behavior according to what they see actual persons doing, or persons figured in images, whether they appear in pictures, including comic books, or on television (Lesser, 25). Years of debate on the subject led to a $1.8 million, 1972 study under the aegis of the Department of Health, Education and Welfare, and published as the U.S. Surgeon General's Report entitled *Television and Growing Up: The Impact of Televised Violence*. The findings were inconclusive, a point that became controversial in itself, but the report both focused the preceding years' debate on the topic and fostered further controversy and study in years ahead, including a 1982 report, *Television and Behavior*, which concludes that after another decade of research, "the consensus among most of the research community is that violence on television does lead to aggressive behavior by children and teenagers who watch the programs" (Volume I, 6; see Cater and Strickland). The vast social research on this still-unresolved question of the relation of television to children's behavior has made its way into journalism constantly since the early 1950s, with the psychologists and sociologists constantly identified as Experts. Again, a small sampling of titles is revealing: "Violence on TV: Entertainment or Menace," "Does Television Condition Children to Violence?," "What TV Violence Can Do to Your Child," "Five Acts of Violence Per Hour," "How On-Screen Violence Hurts Your Kids," "Video Violence" (*Cosmopolitan*, Feb., 1953; *Science Digest*, April, 1962; *Look*, Oct. 22, 1963; *Saturday Review of Literature*, March 14, 1970; *America*, June, 1974; *Redbook*, Nov., 1987; *Essence*, June, 1988).

The struggle inscribed here is a contest between the peaceful parental home and the violent TV, under whose influence the children, the Innocents, become potential insurrectionists, at some point to turn against their peaceful families, their society, their friends, themselves, to expose the monumental failure of the Guardians and repudiate the American ideal of the well-rounded child. *That* child, like "May's child," is social, involved in activities, gregarious, active—and high school yearbook captions to this day measure these attributes in printed lines accompanying the photograph of the graduating senior. Television–the–Corruptor threatens to incite violence against The Home and the Guardians, at the least to steal the child's psyche: "Some children can become so preoccupied with television that they are oblivious to the real world around them" (Cross, 221).

The powerful grip of the allegory is such that relatively few of the Experts challenge the idealized terms in which The Home or the Guardians are cast. The few dissenting voices are useful to hear because, by refusing participation in the allegorical terms, they disclose them, in effect, as allegorical. One recent child-rearing text, its author British, remarks that "The truth is probably that a violent *society* produces violent people *and* screens violent entertainment" (Leach, 677). Benjamin Spock takes that view in his most recent child care text, and it is useful to recall that Dr. Spock, who in the 1960s–1970s was outspoken in his opposition to the Vietnam War, drew severe censure from conservatives who blamed student protest against the war on Spock's permissive child-rearing advice to a generation of parents. Dr. Spock, in this sense, was positioned as an errant American, and as a de novo outsider he suggests that American history and the family are broadly implicated in the subject of children and TV violence:

> Our country is the most violent in the world in murders within the family, rape, wife abuse, and child abuse. Domestic violence is both an expression and a cause of the seething tensions in many of our families. Ours has always been a rough society, slipping easily into brutality, as in our treatment of Native Americans, slaves, and each new wave of immigrants. Now brutality on television and in movies is multiplying the violence. . . . It's not that a child brought up by kind parents will turn into a thug. But everyone is being edged in that direction to a greater or lesser degree. (11)

In venturing to characterize The Home and the nation as violent and brutal, Spock really violates the terms of the allegory, which depends on the premise of the home as peaceable kingdom, and the pre-TV era as an ideal of love and care between the Guardians and the Innocents. Spock, in changing the terms, redefines The Home and the Guardians in such a way that television cannot be the sole corrupter of the Innocents. So, too, are the home and the family, configured in terms that strip them of their traditional connotations.

Spock's passage, however, is unusual, and the question this ongoing allegory poses about the social construction of the child and television is, Why is there no outcome, no resolution, no end? In John Bunyan's *Pilgrim's Progress*, the pilgrim, Christian, journeys through life, struggles, backslides, moves forward, and at last envisions the heavenly city. But that pilgrim had Bunyan to shape the narrative and provide closure, while the allegory of the child and television continues to be written and rewritten in public discourse. In that discourse, the child never grows up, the television continues to be metamorphic and dangerous, and the Guardians' anxiety is ratcheted ever higher.

In fact, although the child is the ostensible subject of this discourse, these texts position the parental Guardians at the center. In their problematic zone, the parents are the real subjects, set in ambivalence, ambiguity, vaguely threatened about their task as TV disciplinarians. And the parents' position vis-à-vis their children and television is untenable. They are simultaneously powerful but disempowered. The parent is the figure held responsible, yet acknowledged to lack control. "Parents usually have some idea as to which books and movies they do not want their young children to read or watch," and yet "television renders parents impotent to control the information flow to their children" (Elkind, 73). These Guardians have good judgment but are "impotent" in its exercise.

Spock's *Dr. Spock on Parenting* (1988) participates in this parental miasma. "It's certainly a problem," he says of the parents who have "high standards" for their children and thus by definition need to limit their TV watching. His text speaks on behalf of these parents: "In a general way, they feel that all the violence and vulgarity is bad, and they don't like the idea of their children sitting for hours and passively accepting entertainment instead of inventing their own play." Spock, like the parents, knows the "fact" that their children "crave" television and that "most other children are allowed to watch to their heart's content" (these presumably the offspring of parents who have 'low standards' for them), and that therefore, "parents don't know where to set the limits." They swing from strictness to leniency and then pay the price: "hesitancy, guilt, and vacillation are noticed instantly by children, and this encourages them to argue and push" (177–78). As guardians, these parents are caught in the untenable position of the empowered who are disempowered. If they yield to the "craving," they violate their own principles, with which Spock says parenthetically that he agrees; but if they do not, they meet their children's resistance in argument and aggression.

There is a normative, prescriptive tone toward the Guardians in these texts, whose statements and rhetorical questions blame and castigate parents for setting bad examples (i.e., behaving like TV children themselves), for neglecting to provide stimulating activities for their offspring, for failing to unplug the television: "Parents should show their children that reading is both enjoyable and useful. They shouldn't spend all their time in front of the TV. . . . Are you addicted to soap operas? Do you look forward to Friday nights so you can space out?" (Cutright, 100, 119). "Would [your daughter] insist on watching right up to bedtime if her father was waiting to read the next installment of their bedtime story?" (Leach, 676). "Think about your child's study habits. . . . Does he do his homework in front of the TV?" (Cutright, 75, 129–30). "The amount [of TV watching] should be restricted and care taken about what is watched. Be sure that you and your child don't automatically turn on TV when you get home and leave it on until bedtime, so that it is background noise in your home" (Cutright, 114–15). "The set is *never* switched, or left, on 'to see what's on' or because 'I'm bored' or as background 'company.'" (Leach, 677). These accusations, recriminations, and exhortations accumulate. Repeatedly, the sins of the TV Innocents are laid at the feet of the powerful but impotent, the responsible but irresponsible Guardians.

There is one apparent scenario of conciliation in the allegory, in which teachers, and by extension, parents, are urged to talk to children about TV programs because "it's the easiest way to wire-tap their private world" (Glasser, 58). If communication between adults and children seems at issue, the verb, "wire-tap," says that surveillance of the Innocents is the real agenda. And this comes up in Spock, too, when the pediatrician reports an article by two parents describing a way of communicating with their children: "They watched some of their children's television programs with them, especially on Saturday mornings." The parents could thus exert authority ("veto the brutal programs") and be authoritative ("the children took the occasion to ask many questions about matters that confused them, and the parents enjoyed being able to enlighten them"). But best of all, says Spock, "this was a pleasant, easy way for the whole family to be friends" (56). Here we return to the television as electronic hearth, but more than that, to the parents reasserting their roles as authoritative Guardians of the Innocents. The scene is one of familial harmony precisely because the parents, not the television, are in control. The parents infiltrate the child–TV world for purposes of surveillance and, in the guise of friendship, exert their guardianship. For the moment, the Innocents are protected.

While public discourse in the United States continues the allegory of the TV child, a popular novel-turned-movie, available for rental under "Comedy" in video stores, has presented a criticism so devastating that it has gone unrecognized. To spend a couple of hours with Jerzy Kosinski's *Being There*, either in novel or film form, is to see that author having a last laugh on the allegorical struggles of American parents, educators, and child-care experts over the problem of television and children. Kosinski's *Being There* (1970), produced as a 1979 motion picture starring Peter Sellers, Shirley MacLaine, and Melvyn Douglas, has been widely read as a criticism of television and its influence in contemporary America. In part it is—but only a fraction of Kosinski's savage satire on American culture had been grasped. The rest ought to be, for it lies at the heart of the question of children and television in the United States.

The allegory of the child and television requires, to state the obvious, the two figures, the child and the adult. Yet Kosinski argues that this nation only produces children. *Being There* deliberately flattens all the generations into one quintessential figure—the child. The novel (and film, which follows it very closely) interests us here as a political fable about the child mind of this nation. Educators, child-care experts, and others have sought to position parents, children, and television in developmental terms emphatic of civilized values and rejecting of the disruption and danger perceived in programs exhibiting violent and antisocial norms. But the Polish novelist Kosinski (b.

1933) who arrived in the United States in 1957, brought a very different point of view to bear on American culture and its relation to television.

Being There is essentially a political fable indicting the United States as a puerile nation. A simple-minded orphan of uncertain parentage, named Chance, has been sheltered from childhood within the walls of an estate, where he tends the garden and otherwise watches television in his room. He is illiterate and unable to write. His life experiences are confined to gardening and observing behavior and speech patterns on television. "The figure on the TV screen looked like his own reflection in a mirror . . . he resembled the man on TV more than he differed from him" (6). Lifelong, servants have brought his meals, and in adulthood he wears the elegant, cast-off clothing of the estate owner, identified only as the Old Man. Chance is, Kosinski suggests, rather like the plants he tends, bereft of higher-order reasoning, powerless at self-reflection. When not tending the garden, he "sank into the screen . . . and Chance, like a TV image, floated into the world" (6).

Kosinski acknowledges that television and its viewer form an interpenetrant unit. But the world of *Being There* offers no profound televised instruction signaling educational worth. There is no McCarthy hearing or Eichmann trial. Rosen's Hattie, we recall, is a selective viewer of taste and judgment who turns off the set at the conclusion of each day's trial episode. She brushes aside a neighbor's suggestion that she seek serious issues in the soap operas instead of the Holocaust. She separates TV gold from dross.

But that very power of discernment—rather, the lack of it—troubles Kosinski. What about the American bred on TV watching and unable to be selective? Chance watches nonstop, and the programs are all one to him, ads and newscasts, situation comedies and political speeches. He is entirely a creature of television. "The figure on the screen looked like his own reflection in a mirror. . . . By changing the channel he could change himself." He is a composite of every TV image he has ever seen. He is the video plastic man. He has no existence beyond the world of television. In pain, his mind blanks, "like a TV suddenly switched off." He and the TV world are one. When at last he goes outside the walls of the Old Man's estate, Chance sees a world resembling what he had seen on TV. "He had the feeling he had seen it all" (6, 28, 29).

The American political and financial establishment is much the butt of Kosinski's satire. For when the Old Man dies and the estate is closed for probate, Chance is cast out of the garden. East of Eden, so to speak, he encounters a society functioning solely on images, one in which individuals— the rich and powerful, the well-connected from diplomats to kingmakers and the President himself—live on nostrums and images, projecting their yearnings and presumptions onto Chance.

Thus, following a minor accident with a chauffeured car, Chance is swept up into the world of wealth and high-level political and economic power as he recuperates, by invitation, at the estate of the financier–owner and his wife. They presume that the syllables of his name, Chance–the–gardener, must really be Chauncy Gardiner. His elegant clothing signifies his place in their class. The story is a semiotic circus, for every character writes his or her version of the Chance image, based on presumptions of signification. His financier–host, a counselor to presidents, presumes Chance's business acumen, while his romantic wife feels certain the man has lost a woman's love. When the president visits the estate, Chance is presumed by the press and later by foreign diplomats to be a high-ranking policy advisor, the "architect" of the president's subsequent speech. The Soviet U.N. ambassador decides that Chance knows the Russian language and literature and that he is playing foreign policy like a chess master. The gardener is offered a book contract, beseiged by the news media for interviews, and actually appears, with great success, on a nighttime talk show, where he performs as an image of all the talk-show images he has watched.

Throughout, everybody mistakes Chance's slow speech and gardening nostrums for self-possession, confidence, and profundity. "In a garden," he says to the attentive, respectful president of the United States, "growth has its season. . . . As long as the roots are not severed, all is well and will be well" (54, 57). The president is inspired. Within four days of leaving his walled garden, his trowel and sprinkler, Chance is on his way to nomination for vice-president of the United States. His lack of a past is considered an asset.

Two related critical issues undergird *Being There*. One concerns the vacuous organic garden figures that can be mistaken for profundity and sophistication in a modern world Kosinski says is fractured by record-breaking unemployment, large-scale business failures, the dangers of Cold War escalation into nuclear Armageddon, Eastern European and Soviet antisemitism, and current warfare (presumably the Vietnam War, given the publication date of the novel). It is unsurprising that the European, Kosinski, should grasp the political and socioeconomic danger in this shockingly naive and simplistic rhetoric of the American ruling class of the later twentieth century. "In this colored world of television," Kosinski says, "gardening was the white cane of a blind man" (5). (And in a stunning validation of Kosinski's critique, one of those moments in which life imitates art, we find President Ronald Reagan's former Budget Director, David Stockman, recounting a meeting between the president and the secretary of the treasury, Donald Regan, held on November 2, 1981, a year of severe recession. The president, Stockman reports, was informed that his economic program of tax cuts was a pipe dream, that if enacted, it actually would lead to a one to two billion-dollar deficit over the next few years. The president, taken aback, was reportedly most "soothed" by

Treasury Secretary Regan's dismissal of dire projections. "Let me give you a homely analogy," said Regan to the president. "It's as though speculators in grain were to say that we're going to have a crop failure a month after the farmer has planted his seeds. How could they know that? Anything can happen. Let's give these seeds we've planted a chance to grow" [Stockman, 373]).

Like numerous Europeans, Kosinski is contemptuous of America's ahistoricism. Absorbed in images in the present moment, the nation lacks all historical depth, all of the dimensionality of human processes through time. Fixed on the present instant, the nation lacks the historical imagination. In *Being There*, television is no vehicle for the revivification of history, as it is in Hattie's world, but on the contrary is the most antihistorical of media, engaging the viewer—in fact, the entire society—in a continuous present without reference to past or future:

> Of all the manifold things there were in all the world—trees, grass, flowers, telephones, radios, elevators—only TV constantly held up a mirror to its own neither solid nor fluid face. . . . Television reflected only people's surfaces; it kept peeling their images from their bodies until they were sucked into the caverns of their viewers' eyes, forever beyond retrieval, to disappear. (63–65)

Television, for Kosinski, is traitor to all civilized values, and his novel argues that the society that lives by television exists, by definition, in a state of arrested development. Chance is deliberately infantalized by the novelist, and not only to become a blank slate onto which everybody can project qualities they are predisposed to find there. More than that, Chance is the quintessential childlike American. Kosinski carefully makes him asexual, forever prepubescent as he is forever simple-minded. He represents the pejorative associations of the very word, child. Through Chance, Kosinski argues that a television culture is another term for an infantile culture. There are no adults here in the United States, says Kosinski, only TV children in a state of arrested development, children in disguise.

Comics, Movies, Music, Stories, Art, TV-on-TV, Etc.

Television, its advertising, and the popular culture they both reflect and define have fundamentally altered what intellectuals get to regard as the proper objects of their attention.

David Foster Wallace, "Fictional Futures and the Conspicuously Young," 1988

Put yourself behind a Pepsi . . . You've got a lot to live, and Pepsi's got a lot to give! The Pepsi-Cola Company hopes you have enjoyed our ad with tightly edited images of happy people having far more fun than you ever do! Now back to our program.

G. B. Trudeau, *Dare to Be Great, Ms. Caucus: A Doonesbury Book,* 1975

One way filthy rich TV game show producers become even more filthy rich is by spinning their shows off into home-version board games. We think this is outrageous! . . . So we're offering . . . Mad's home version of "Jeopardy!" . . . [300-point answer:] The Full Names of the Three Stooges. . . . [question:] What are Dan Rather, Peter Jennings and Tom Brokaw?

J. Prete, *Mad* Magazine, 1987

"And now, kids, it's time for 'Mister Rogers' Neighborhood!'
"Hello Boys and Girls! Welcome to my neighborhood! Won't you be my neighbor?
"Good! OK boys and girls, let's go to the magic kingdom and meet our guest adults for today . . . Mr. Jones. He's a tuba player. Can you say 'tuba player?'
"Tuphlem Grdlphump."
"Good!"

Berke Breathed, *Bloom Country Babylon,* 1986

Well I dreamed there was an island
That rose up from the sea.
And everybody on the island
Was somebody from TV.
And there was a beautiful view
But nobody could see.
Cause everybody on the island
Was saying: Look at me! Look at Me!
Look at Me! Look at me!
Lauri Anderson, "Love Is a Virus"

The TV environment ratifies itself everywhere. Cartoons, comic strips, fabric prints, sculpture, music, paintings, flip books, T-shirts, jewelry, movies, and TV itself—these, along with printed texts, have featured television prominently, often critically, both attacking television and at the same time exploiting its resources, but above all affirming and validating the TV environment. Television is by now ubiquitous in virtually every cultural format and venue in the United States. It takes shape as familial hearth, as the illuminator/corruptor of children, as the paradoxical site of sedentary activism, as the locus of a new, multivalent consciousness. It is a source of language, virtually a contemporary phrasebook, and certifies human experience in contexts ranging from sports stadiums to personal spaces where camcorder cassette tapes are played on personal screens. Every sign of it, from a T-shirt front to a refrigerator magnet reinforces the idea of the TV environment, one extending from the Magic Screen on "Pee-Wee's Playhouse" to the video apparatus (videocamera, VCR, big-screen monitor) on which the pantomime, Will Irwin, the electronic-age Charlie Chaplin, performs onstage in his one-man video vaudeville act. Everywhere television is ratified as it is reified in contemporary culture.

Venues

Rock music exploits and validates television, even in critical terms. "I'm a television man," sing the Talking Heads, the group's name itself a TV reference. "The Sun Only Shines on TV," sings the group, a-ha, suggesting the idealization of life by television, though other groups enumerate relentless disturbing TV messages. The B-52s' "Channel Z" tells of the terrifying global "static" imaged on television, including environmental damage, nuclear radiation, stock market crashes, together with fast-food ads and politicians as a drug cartel. The song is an exhortation to break free of channel (A-to-) "Z" static—but also an acknowledgment that television is the epistemic orientation

Costume jewelry pin in TV motif, 1990.

of contemporary life. The Talking Heads' "Television Man" describes consciousness itself in these TV terms, arguing that contemporary consciousness is structured and governed by television. "The world crashes into my living room. . . . I'm inside and I'm outside at the same time."

Rock is not the only music exploiting television. In New York City, in 1989, the composer John Moran, a protege of Philip Glass, presented an opera, "The Jack Benny Program," comprised of videotape recordings of "The Jack Benny Show," which was broadcast for fifteen years, 1950–1965, the so-called golden age of television. "Parts of the work include skits from the show, but for long sketches, Moran has used time-honored electronic music techniques to transform the material into a more abstract stream of sounds and textures." The composer's collaborator explains the basis of the opera: "To me, the television shows of the 1950s and 1960s are our iconography" (Kozinn).

Movies, too, have incorporated television, from the 1955 Hollywood film, *All That Heaven Allows*, which shows a TV set presented to a lonely woman (Jane Wyman) as a consumerist replacement for love and companionship (see Joyrich), down to movies like *Network* (1976) and *Broadcast News* (1987), whose titles speak for themselves as they purport to expose the actualities of TV

Boston's Bull-and-Finch Pub at the Hampshire House, best known as TV's "Cheers" bar and, on that basis, a popular Boston tourist site. Photo by Bill Tichi.

T-shirt with television imprint.

Motivational poster, 1989–1990.

news. *Capricorn One* (1978), with Peter Hyams, Elliott Gould, and O. J. Simpson, is based on the massive hoax of the first manned flight to Mars, a hoax instigated by the CIA, NASA, and government officials but perpetrated through TV news organizations. The 1986 *True Stories* argues the pervasiveness of television, for screens are everywhere, in church, in a private shrine, in the roadhouse bar, in the very architectural shape of an outdoor talent-show stage—and most of all, in the bedroom, in which a character called "The Lazy Woman" (Swoosie Kurtz) lives her entire life in bed in front of the television and, using it as a matrimonial shopping channel, finds her man, Lewis Fyne (John Goodman), who in the closing scene marries her in a bedroom wedding as guests mingle. The two will live happily ever after in front of television. The front is not enough elsewhere, however, for another film, the horror feature, *Shocker* (1989), sends characters into the receiver chassis itself, raising TV environmentalism to new literalism in its climactic chase scenes. In 1990, the movies have begun to narrate the history of television in domestic and commercial life. The Barry Levinson film, *Avalon*, shows the postwar family gathered around the new TV set watching the fixed image of the test pattern. Throughout the film, a multigenerational family chronicle, the television programs of the late 1940s–1970s air in the background, and moviegoers see a capsule history of television cabinet design from console to portable, not only in the habitat but in the appliance store operated by two cousins.

And daily the comics are saturated with television. Newspaper and magazine cartoonists and comic strip artists, like painters and sculptors, were responding to television in the 1950s–1960s, for instance in Henry Scarpelli's widely syndicated TV TEE-HEES or Ted Keys's "Hazel" series in *Look Magazine*, in which the uniformed maid made acerbic comments on family life and child-rearing. And for years we have seen the satiric exploitation of particular TV programs in *Mad* magazine, which regularly satirizes popular shows like "Miami Vice" or "Murder She Wrote"—or "The Cosby Show," in which the son, Theo (called "Zero" by *Mad*), asks his father, the Doctor Huxtable character, "If we're *Black*, how come we're so well off?!? How come we *act* White . . . and THINK White??" The Bill Cosby figure replies, *"That's* so the *White viewers* can *identify* with us! How *ELSE* do you think we get such *HIGH RATINGS*?? (*Mad*, June, 1985, pp. 44–47).

Current cartoonists like Tom Wilson ("Ziggy") and Gary Larson ("The Far Side") also utilize television for satiric purposes, Wilson equating television's Pee-Wee Herman with Vice-President Dan Quayle, Larson suggesting that the pre-TV era family was just as enervated as the "boob tube" family is made out to be, as his "in-the-days-before-television" family sprawl in their living room, staring ahead at a blank wall, their dog the most alert figure among them. Daily, the comics exploit the TV set as domestic furniture and satiric opportunity, for instance in Cathy Guisewite's strip, "Cathy," or Doug Marlette's "Kudzu," which regularly shows the greedy, hypocritical tele-vangelist, the Reverend Will B. Dunn, on-screen making his pitch.

As for television itself, traditions of using TV-on-TV have become so well established that MTV's quiz program, "Remote Control," can operate entirely within a self-referential world of television, with the three contestants in their TV recliners, choosing question categories drawn solely from TV with hand-held remotes. But television has exploited itself at least since the 1950s and 1960s, when "The Jack Benny Show" based itself on Jack's problems with his weekly television program, and "The George Burns and Gracie Allen Show" featured the pixilated Gracie performing inside a TV cabinet and her husband George spoofing TV conventions by observing on-screen action and making asides directly to the viewing audience. Television has continued this reifica-tion process, for instance with the 1970s comedy series, "The Mary Tyler Moore Show," followed by "WKRP," "Murphy Brown," and "WIOU," all purporting to show behind-the-scenes work life at TV stations. Certain programs, moreover, have lent themselves to repeated TV satire. Over the years "Saturday Night Live" has done skits satirizing "Sixty Minutes," the talk show, "Attitudes" and "thirtysomething," and Eddie Murphy has lampooned "Mister Rogers' Neighborhood" with the white middle-class suburb recast as an urban black ghetto. Murphy, like Fred Rogers, puts on his cardigan sweater and changes shoes, but tells the "children" that his wife has left him ("the bitch"), and answers his front door as if fending off a drug bust.

And in art, painters are incorporating television in their compositions, for instance, May Stevens's *Prime Time* (1967), one of her Big Daddy series condemning authoritarian patriarchy, or Andy Warhol's $199 *Television* (1960) or Robert Watts's *TV Dinner* (1965). The painter Janet Fish, whose work recurrently has involved the formal problem of rendering light through transparent materials, includes the glass-screened television (with a football game shown in progress) along with colored glass, pliofilm, plastics, cello-phane in her *Hungry Man Dinner* (1983). Another artist, Jan Havens, suggests that TV news is the format for the reconception of major religiomythic events. Accordingly, Havens shows a television on which a boyish T-shirted God in the foreground is interviewed by a TV reporter–angel at the scene of the Flood and asked for His explanation of the water, the aninals, the Ark floating in the televised background (*God Explaining the Flood*, 1987). And video art necessarily flourishes in counterpoint to broadcast television, as seen in the Korean-born, New York-based Nam June Paik's TV sculptures and installa-tions featured in numerous exhibitions, among them the 1989–1990 "Image World: Art and Media Culture," at New York's Whitney Museum of Ameri-can Art, which also featured the video art of Bill Viola and Frank Gillette.

Janet Fish, *Hungry Man Dinner,* shows the legitimation of TV as the artist's subject matter. Courtesy, Robert Miller Gallery, New York.

Nashville-based artist, Jan Havens, suggests in her 1989 "God Explaining the Flood" that television is the contemporary venue for the representation of reli-giomythic events. Reprinted by permission of the artist.

And fiction, like other arts, exploits television for form and cultural reference, as we have noticed throughout this discussion. Certain recent fiction texts directly engage television and are instructive because they show both a critical practice and an immersion in television, presuming utter familiarity with television, its programs, its reruns. Donald Barthelme's "And Now Let's Hear It for the Ed Sullivan Show" (1974) seems to read like a transcript of that variety hour, which began in 1948 as Ed Sullivan's "Toast of the Town," becoming the "Ed Sullivan Show" in 1955 and running in Sunday prime time altogether for twenty-three years. Barthelme's story follows the format of a typical program. "The Ed Sullivan Show," he begins, "Sunday night, Church of the unchurched. Ed stands there. He looks great" (102). Like a transcript, the story moves through the typical melange of Sullivan show acts from a singing group, Doodletown Pipers, to an inspirational recitation by Helen Hayes—except for one parenthetical digression on a pornographic scene that, the narrator tells us, mysteriously appeared on a Palm Springs TV station, "then disappeared into the history of electricity." He concludes, "What we really want in this world, we can't have" (107). But Barthelme's one line repositions the entire story. Suddenly the reader is made to understand that the show, for that matter every such show, is an evocation of unattainable yearnings and tantalizations, that the basis of television is its perpetual promise of an unattainable fulfillment.

And T. Coraghessan Boyle's "Heart of a Champion" (1975) operates from the same premise of reader familiarity in its reprise of "Lassie," which one commentator calls "one of the most durable properties on television," begun in 1954 and carried on cable stations as reruns, as in the early 1990s yet another series moves into production. The staple of "Lassie" is the collie's unfailing rescue of those endangered, including strangers and members of one of the two casts of family characters in the series, with Tommy Rettig as the original Timmy, and, as the mothers, Cloris Leachman and June Lockhart (Marschall, 107). Boyle opens his story in the idealized rural setting of farm fields that opened every TV episode of "Lassie," though when we hear of Timmy's eyes "blue as tinted lenses," we know Boyle's "Lassie" is veering sharply into irony even as the boy whistles for the dog.

The black-and-white TV series is now colorized for fiction, for Boyle has in mind a satiric, subversive "Lassie." And sure enough, the collie is a contraption ("fast as a flashcube . . . paws churning like pistons"), and Boyle's conflation of the story lines familiar to nearly four decades of viewers exposes the TV fantasy of the dog-hero. Boyle's Lassie rescues the imperiled with mouth-to-mouth resuscitation and with winch-and-cable devices and a chain saw, and she barks English sentences ("Yarf! Yarfata-yarf-yarf! Arpit-arp"—translation: "Timmy's trapped under a pine tree out by the old Indian burial ground, a mile and a half past the north pasture" [405–7]. Boyle exposes the TV sentimentality of canine domestication too, when Lassie mates with a syphilitic coyote who later attacks Timmy and settles down to eat the boy's hand. Lassie, at first leaping to help, drops her "show-dog indignation" and shows her evolutionary savagery by joining the feast.

The satiric butt of "Heart of a Champion" is both the TV series and the TV audience who are at this moment also the readers. This story depends entirely on a reader knowledgeable about the series, its conventions, its story lines, its suppressions. Readers are forced now to recognize the sentimentality of the series and the complicity of viewers with its norms. Our very nostalgia is presumed, exposed, doubled back on itself.

"Heart of a Champion" broaches the fictional hyperreal, and as a hyperrealist insider, another writer, David Foster Wallace presents TV fictions based on "Jeopardy" and "The David Letterman Show," and also on the on-screen presidents of the television era. Wallace's reader, like Boyle's, needs to know the TV programs to know the fiction. Here is a passage from Wallace's "Little Expressionless Animals," in which Alex Trebek, the game show host of

1989 Hallmark card presuming widespread public knowledge of TV's
"Lassie." Reprinted courtesy of Hallmark Cards, Inc.

television's "Jeopardy" finishes a racquetball session with his predecessor on
the show, Pat Sajak:

> Pat Sajak comes close to skunking Alex Trebek in three straight games of
> racquetball. In the health club's locker room Trebek experiments with a half-
> Windsor and congratulates Sajak on the contract renewal and iterates hopes for
> no hard feelings re that Applause-sign gag, still. Sajak says he's forgotten all
> about it, and calls Trebek big fella; and there's some towel-snapping and
> general camaraderie. (*Girl with Curious Hair*, 37)

Later in the story, we hear that Alex Trebek buys a car "so expensive he's worried about driving it. He takes the bus to work."

Temptation is, to read passages like these as a spoof on synthetic celebrities or as intentional fictional titillation. But this is not the burst balloon of the reality conjectured and disclosed behind the on-camera facade. Wallace, instead, is working out of a newer realm of the hyperreal, of the authentic achieved through the absolutely fake, as Umberto Eco put it. Thus his stories, "Little Expressionless Animals" based in the realm of *Jeopardy*, or "Lyndon" on the Johnson presidency, or "My Appearance," in which a well-known actress prepares to appear on the "David Letterman Show" and is coached in being "not sincere" but not insincere by her TV-producer husband. Wallace's disclaimer at the outset of *Girl with Curious Hair* is revealing: "Some [of the stories] project the names of 'real' public figures onto made-up characters in made-up circumstances. Where the names of corporate, media, or political figures are used here, those names are meant only to denote figures, images, the stuff of collective dreams; they do not denote, or pretend to private information about, actual 3-D persons." These are projections, not three-dimensional persons but fictional holograms. The absolute fake of the projection is the authenticity of the fiction. Wallace is the TV insider as hyperrealist.

And so is Mark Leyner, whose *My Cousin, My Gastroenterologist* (1990) is blatantly hyperreal fiction self-evidently indebted to television:

> I am on every channel and that infuriates you
> that I have the ability to jump out of the television screen,
>   burrow into your uterus, and emerge nine months later
>   tan and rested bugs you very much
> you're using the violent vocabulary of the u.s.a., you're
>   violently chewing your cheez doodles and flicking the
>   remote control
> a computer programmer from bethesda, maryland, puts her
>   fingers through the holes in my head and bowls me
> i am rolling through roanoke, city of rheumatism and alzheimer's dis-
>   ease. (99)

Disjunctive, flaunting discontinuity and simulation, this kind of writing evidently finds an audience who welcomes the print text that can ratify the TV-age cognitive reality. One commentator compares Leyner's style to MTV, and a reader remarks, "It's the whole experience, of watching TV and being in the room and the cars going by. It's much more true to life than a [traditional] narrative" (Hall).

All these manifestations of TV argue the process of acculturation brought to extremes. The passages quoted above obviously depend on television for their meaning and presume reliance on audience familiarity with the TV references. These diverse venues from fiction to painting show a TV environment so thoroughly established that it is a resource exploitable for form, for technique, for the coding of messages sent and received in the artist–audience relationship.

Insiders/Outsiders

Caution is necessary here. Not every artist, cartoonist, film maker, TV producer, or writer who puts a TV set in a production, illustration, art text, and so on necessarily works from *within* the traits of the medium even if that person seeks to purvey an insider's version of television. The point seems obvious, its implications perhaps less so. Artists and others can only enact the values of a video culture from their cognitive presence *within* the TV environment. Those psychologically or cognitively outside of it, sometimes but not always those born in the years before the TV consoles, table models, and portables proliferated in American households and public places, maintain a very different relation to the medium even when they exploit it in their work.

At issue here is the *psychological* enclosure within or outside of television on the part of the text and of the audience, whether the text is a movie, a video, a comic strip, and whether the audience is sitting in a cinema complex, strolling the space of an art gallery, sitting at breakfast with a newspaper, listening to rock music lyrics. The crucial question about the TV text is whether it is enacted from a place psychologically inside or outside the television environment.

Certain print texts, not surprisingly, show this insider/outsider relation to television, and because the issue is so important in the portrayal of television in film, comics, videos, a pairing of passages in novels is helpful, one from Bobbie Ann Mason's *In Country*, adapted into film in 1989, the other from John Updike's *Roger's Version*, which depicts the household of a Protestant theologian, Roger Lambert, who watches his son "crouching blurry-eyed over his math homework while trying to keep a rerun of 'Gilligan's Island' in focus":

> On "Gilligan's Island" a small man with a yelping voice was wearing a sarong and trying to avoid a heavyset blond man who, clad in a splashy-patterned bathing suit, was bombarding him with water balloons from a helicopter. . . . "Gilligan's Island" momentarily yielded to a commercial. For catfood. A handsome, caramel-colored cat, an actor–cat wearing a bow tie, was shown snubbing raw steak and fresh fish and then greedily burying its face up to its throat muff in a dish of gray-brown pellets. (44–48)

In this passage Updike wants his narrator and reader to be critically detached from the everyday culture of television. He does not see Gilligan as a character, much less as the actor Bob Denver, but instead as a small man with a yelping voice in a sarong. Watching the commercial, he does not see Morris the cat but an "actor–cat." Updike intends that his narrator and his reader be dignified by a critical distance from the TV environment. We are supposed to be above and beyond all that. Momentary analysts of the TV rerun and the commercial, we remain superior in our distance. Our very ignorance of the specifics of the TV names is important. To name names, to say "Gilligan" or "Morris" is to implicate ourselves in the crass, the commercial TV culture. Our only correct response is a detachment that means rejection of the television world. Both Gilligan and Morris are American bywords, but Updike begs to conspire with the reader not to say what he knows, or even *that* he knows. Bringing television into the novel, Updike works to distance himself and his proper reader from it.

In contrast, Bobbie Ann Mason's novel, *In Country* (1985) repeatedly places the reader before the TV set in direct engagement. In one scene from this post-Vietnam antiwar novel we see eighteen-year-old Samantha (Sam) Hughes in a motel room with her uncle Emmett watching the "Tonight Show" in which the comedian Joan Rivers substitutes for Carson:

> Joan Rivers is wearing a black taffeta job with a balloon ruffle around the hips and gobs of pearls. She says her outfit is Boy George's gym suit. She is made up pretty and blond, but she isn't really that pretty or that blond. She says it's raining in L.A.—at last Willie Nelson will get his hair washed, she says. She says he wears a Roach Motel around his neck. Her first guest is Don Rickles. Don Rickles tells Joan Rivers, "Johnny hired you because you're no threat." He says Johnny is at home, posing in his swim suit and saying, "How's the body?" Don Rickles and Joan Rivers rattle back and forth about their dates in Las Vegas. Joan Rivers says a woman needs a funny face and a trick pelvis and that's all. But Don Rickles says college would be an advantage. His daughter is going to college. (19–20)

The narrative consciousness, Sam's, is directly involved in the televised segment, which is virtually transcribed, something of a documentary report, though not without criticism. Rivers, Sam observes, is "made up" to look pretty but "isn't really that pretty," and she and her show business guest "rattle" on about their forthcoming appearances. Readers will soon understand that the decision to go to college hangs heavily on Sam just now, so the Don Rickles statement about his daughter has a personal immediacy and importance.

This text, unlike Updike's, does not insinuate two classes of readers, those seduced by television and those who stand above it. Mason nowhere signals her own superiority to television. She does not patronize or condescend to her characters. Nor does she invoke a tone of ironic ignorance to provide her readers an exit to an Olympian critical distance from the televised scene. We hear nothing, for instance, of a self-deprecating, angular blond stand-up comic who insults other celebrities by satirizing their physical appearance.

Quite the reverse. The text presumes an audience who are TV insiders. No distinction between the mature and the puerile hinges on knowledge or ignorance of commercial television. Mason is *within* the TV environment and presumes that her reader is also there.

Art texts also exemplify this distinction—in this case a sculpture and a line drawing, Edward Kienholz's *Six o'Clock News* (1964) and Saul Steinberg's untitled drawing of a TV woman sitting on a man's lap, from *The Inspector* (1973). These show radically different relations of the artist to television. Although both directly exploit television, their relations to the subject are so different as to be antithetical.

Kienholz's *Six o'Clock News* is a sculptural construction of a TV set fashioned from a rusted metal container with large knobs and the indoor antenna called rabbit ears. The news anchor shown on-screen is a toy Mickey Mouse. The sculpture argues that the news is trivialized, that it functions at the level of cartoon entertainment, that the so-called TV journalist on-screen is nothing more than a corporate logo, as Mickey Mouse is to the Disney Corporation, and that the audience watching in uncritical absorption is mentally no more than children themselves (see Stich, 120–21). The sculpture urges its audience to join with the artist in deploring this state of American culture. It also distances the artist and the viewer of *Six o'Clock News* from television itself. The artist has repudiated it, as must the audience. There is a politically correct response to television demanded by this sculpture—to regret and deplore it, and to distance oneself from it. This is a thoroughly antitelevision text.

Saul Steinberg's erotic line drawing of the voluptuous TV woman, on the contrary, works from within an appreciation of the visual tactile quality of the on-screen image. Steinberg exploits a trait of television in ways that are technically advantageous to him. Like Kienholz's *Six o'Clock News*, the drawing does make a statement critical of television, but it also invites the onlooker to appreciate the formal properties of the medium. Steinberg's line drawing reminds us, first, of the many humor cartoons showing men enjoying heterosexual fantasies while watching television. In the fantasy, the sexually provocative woman emerges from the TV set to make love with the male viewer. And this is Steinberg's image: a sexy woman sits on the lap of a man who is utterly respectable in his suit and necktie, betraying no sexual yearnings whatever. He embraces her and, of the two figures, she is the more solid, sculptural form, far more three-dimensional than he. We know her to be a TV fantasy in part because of the electric plug that meanders from her heel to a wall outlet and in part because of the rabbit-ear antenna protruding from her head in a comic V.

But most of all, we recognize this woman as a TV fantasy because of the way

From Saul Steinberg, *The Inspector*, 1973. Reprinted by permission of Saul Steinberg.

she is rendered, in the kinetic horizontal lines characteristic of TV transmission. Steinberg's artistic ally is the technical trait of the TV image that is so often criticized in regard to TV transmission in the United States. That is, the relatively low number of horizontal scan lines that lessens the picture quality by diminishing definition is exactly the aspect of the image that Steinberg utilizes—virtually celebrates—in his drawing. His rendering invites us to appreciate the kinetics, the visually tactile quality of transmission, the constant scanning in the rapid horizontals that give the image its open borders. Steinberg's drawing does argue that American erotics goes on beneath middle-class facades of asexual respectability and that television itself is powerfully seductive and chimerical. The apparent solidity of the TV woman over and against the two-dimensional man suggests the way in which the television image can seem more substantial than oneself. But along with Steinberg's critical statement, there is an exuberant exploitation of the TV image, an appreciation of its traits, which are deployed in the art text. Steinberg presumes our utter familiarity with these scan lines, then defamiliarizes them in his drawing. This is the exploitation of television that positions Steinberg *inside* the TV environment, while in *Six o'Clock News* Kienholz remains outside it.

What, then, can we expect from the TV insider? In art, comics, movies, this insider/outsider distinction is crucial precisely because it is central to the prerogatives of a text to be at once analytically critical of television, even adversarial toward television, but, at the same time, to utilize and exploit the attributes of the medium, knowing that the electronic environment is inescapable. As the various texts reveal, television is an environment to be deconstructed, decoded, exposed and, at times, combated. Many TV texts are webbed in ambivalence toward television. Environmentally, television is both hostile territory and home turf—but its environmentalism is the undergirding assumption. The insider, acknowledging this very status, has far more freedom to range critically over the television culture and, simultaneously, far more technical and tonal versatility in practice. There are assenting insiders and resistant ones, as we see; but the insider subject position is vitally important. The outsider text, which presumes that a stance outside the TV environment is still possible, finds itself under severe restriction of the kind encountered in the sculpture *Six o'Clock News* or the novel *Roger's Version*. If it averts its gaze from the culture of television on grounds of, say, commercialism or vulgarity, it becomes burdened by its dissembled ignorance of television, as Updike shows, pretending not to know the name of a title character yet contradictorily acknowledging that "a rerun of Gilligan's Island" is playing.

If the outsider text attacks television from a premise of detachment, it is confined tonally to anti-TV polemic of the kind recycled in decades of repudiation of television. Thomas Pynchon's *Vineland* (1990) shows this, with the TV viewer as substance-abusing "Tubefreek" needing a detoxification program ("Tubaldetox") in which addicts sing the hymn, "The Tube," with lyrics on "poi-soning your brain," "driving you insane," "shoot-ing rays at you," and "see[ing] you in your bedroom . . . know[ing] your ev'ry thought" (336–37). All the anti-TV antagonism is recycled in the familiar terms of addiction, surveillance, mind-control. This is ritualistic incantation of the kind familiar since the late 1940s. Though the narrative is peppered with names of television programs, Vineland is an outsider text, its terms sounding at best like language drills or doctrinal exercises.

Screen

One TV characteristic dominates all others—the screen itself, by now naturalized not only as an electronic hearth but legitimated as intrinsically environmental. Screens for convenience-store monitoring, screens promoting products in supermarkets and at cosmetics counters, screens in museum exhibitions, screens at airports, not only displaying flight information but weather conditions nationally, globally, screens at sports events and rock concerts and in churches and the rooms of public buildings for overflow audiences or those sitting too far back can get a close-up view of the main event. The screen is the ubiquitous environment.

For the illustrator, it is an open panel. Garry Trudeau, for instance, is often dismayed and outraged by what is broadcast on television, especially American politics. Yet the TV screen itself is his venue for sociocultural criticism, and throughout Doonesbury, Trudeau uses the screen consistently as his critical forum. In over thirty books published since the early 1970s, Trudeau has constantly used the TV screen as the venue for his attacks on sexist advertisements, political "chatter," and nuclear brinksmanship. The TV screen remains, for the Doonesbury strip, a venue for Trudeau's castigation of bellicose businessmen, of self-serving political rivalries that overshadow public interest, of book promotions of ghosted "instant books," of verbose and egocentric talk-show hosts, and of the disconcerting interchangeability of the beloved news reader–anchor and the professional politician (Walter Cronkite for vice-president). Trudeau assumes that television is the common culture that he shares with his readers, and the Doonesbury strip relies regularly on the TV screen itself as an invaluable resource. That screen is Trudeau's drawing tablet, infinitely open to new images and infinitely erasable to make way for new ones. It is the cartoonist's perpetually open panel-within-a-panel (cf. Call Me When You Find America, 10–11, 14–15, 70, 117; Bravo for Life's Little Ironies, n.p.; Guilty, Guilty, Guilty!, 12, 103; You Give Great Meeting, Sid!, n.p.; He's Never Heard of You, Either, n.p.).

IN PEACE-RAVAGED SINAI, THIS IS ROLAND BURTON HEDLEY, JR.

Garry Trudeau's *Doonesbury* makes the TV screen a forum for social criticism, in this instance a barb at the vacuous TV celebrity journalist. Reprinted by permission of Universal Press Syndicate.

Rock videos, too, regularly exploit the possibilities of the TV screen, and viewers of MTV have seen this throughout the 1980s (see Kaplan, *Rocking Around the Clock*). "Cult of Personality," by Living Colour, for instance, opens with a young girl watching television, then intermittently returns to register her anguish as clips of fascism and dictatorship (Mussolini, Stalin, the assassinated Martin Luther King) appear in the video, until, at the end, she is shown reaching to turn off the TV set. Obversely, "Rocket," the Herbie Hancock video, presents the on-screen image of Hancock himself at the keyboard of a synthesizer. The image is the preeminent human reality, for Hancock appears on the screens of several color televisions, the on-screen Hancock asserting the power of the screen image and, by implication, the power of music produced on the synthesizer. And "Welcome to the Jungle," by Guns and Roses, shows young girls watching a half-dozen screens on which the dominant images are of violence and material luxury. These girls represent innocence and youth, and we see one of them bound to a chair and forced to watch the modern "jungle" of materialism and police violence—all projected from the screen environment.

Other videos use the screen to reveal emotion. "Don't You (Forget about Me)," by Simple Minds experiments with on-screen visualization of the feelings stated and evoked by the lyrics, while "The Flame," by Cheap Trick, takes this kind of visualization of feeling a step further, two separate screens of the separated lovers enacting the refrain, "I can't believe you're gone." And Bruce Springsteen's "Glory Days" also exploits the screen to capture a state of mind, in this case the poignance of a mature man's recollections of youthful yearnings to play major league baseball. The baseball clip shown on-screen evokes both past and present mentality of the older Springsteen as a working man and a father who still cherishes his boyhood memories and his dreams.

Mainstream television, too, has begun to grasp the possibilities of the TV screen. "What's Allen Watching," a CBS comedy series pilot, featuring Coren (Corky) Nemec as the title character, exploits the TV screen for multiple uses ranging from the display of the teenage Allen's erotic fantasies to transitions into subplots. In "What's Allen Watching," the suburban high school senior, Allen, lives in the white middle-class nuclear American family (Barbara Barrie and Peter Goetz as the parents) whose grown children cling to home.

Escaping to his room, removed from the household traffic, Allen sinks into his recliner and snaps on his television, where he flips around the channels—and views a variety of comic routines, these cast as commercials and regular programs. In addition, he sees clips of the subplots forthcoming during the evening's program. Here we begin to grasp the uses of Allen's TV screen. Not only does the flip of a channel bring in humorously irreverent sketches, as if "Saturday Night Live" had moved to tape, but Allen, who often faces the camera and talks to viewers directly, also interacts with the on-screen figures. For instance, he begins talking to the showering soap-commercial girl, who offers to show him her lathered nude body of which she is "proud." And the boy's library of videotapes makes it easy to show his past, his school life, the girl for whom he yearns. He interacts with her on-screen image too, and as they talk, we realize that it is Allen's fantasy life appearing on-screen, that his state of mind, his feelings, his yearnings will be televised on the screen in his room. We see, in short, that technically TV-on-TV can now dispense with fade-out transitions into flashback scenes of the past or into atmospherically ineffable dream sequences. TV-on-TV can now break the grip of the proscenium arch and the theatrical "fourth wall." The screen will display the character's inner life directly. And a variety of acts and other material can be presented in the format of channel switching. The very naturalization of television over decades makes possible these new bases for the theatrical suspension of disbelief. In "What's Allen Watching," the floorboards of the stage become the multichannels of cable TV or of the satellite dish, and the master of ceremonies is the boy with the hand-held remote.

And the contemporary fiction writer has been quick to see this kind of potential of the TV screen for his or her own work. An on-screen moment represented in a novel can take the place of the excursion into characters' minds usually signaled by the speech tags, "he thought," "she felt." It can supplant the often awkwardly triggered flashback into past events. The televised scene can reveal new dimensions of the fictional characters' lives directly in the moment.

It is helpful to "view" a fictional televised scene, this from Richard Ford's novel, *The Sportswriter* (1986), in which the protagonist, Frank Bascomb, a sports writer en route to an out-of-town interview, stops overnight in a Detroit motel. His girlfriend, Vicki, is with him, and this scene takes place in the early A.M. as he contemplates his possible future with her, one he feels he would like "as well as it's possible to like any life":

> Vicki turns on the television and takes up a rapt stare at its flickering luminance. It's ice skating at 2 A.M. . . . Austria, by the looks of it. Cinzano and Rolex decorate the boards. Tai [Babilonia] and Randy [Gardner] are skating under steely control. He is Mr. Elegance—flying camels, double Salchows, perfect splits and lofts. She is all in the world a man could want, vulnerable yet fiery, lithe as a swan, in this their once-in-a-lifetime, everything right for a flawless 10. Together they perform a perfect double axel, two soaring triple toe loops, a spinning Luta jump, then come to rest with Tai in a death spiral on the white ice, Randy her goodly knight. And the Austrians cannot control it one more second. These two are as good as the Protopopovs, and they're Americans. Who cares if they missed the Olympics? Who cares if rumors are true that they despise each other? Who cares if Tai is not so beautiful up close (who is, *ever*)? She is still exotic as a Berber with regal thighs and thunderous breasts. And what's important is they have given it their everything, as they always do, and every Austrian wishes he could be an American for just one minute and can't resist feeling right with the world. (140–41)

At this point Frank's girlfriend cries that she wants to be the woman skater, who represents a fantasy life of the "stars." But Frank sees in the televised skaters a glowing version of his imagined yet plausible life with Vicki. The two of them could skate, so to speak, together, being at their best something of what the skating pair represent on-screen. And Frank, who is also stirred by nationalistic pride, sees the ideal of this imagined life, as he says, as "a natural extension of almost all my current attitudes taken out beyond what I now know."

For the novelist, the televised skating scene is an irresistible technical resource that becomes a thematic resource as well. There is no way in which television is a novelist's shortcut; the well-crafted scene serves as one fine example of contemporary fiction enhanced by its recourse to television. The televised scene can reveal new dimensions of the fictional characters' lives directly in the moment. The screen becomes the locus of the bared psyche—as it does in the video or the TV program or the comics. If the on-screen images seem at first unrelated to the fictional scene in progress, readers must understand that the writer positions the two—the images on-screen and off-screen—in a kind of fictional haiku, in which two seemingly unrelated sets of images are juxtaposed, the reader challenged to discover their apposition. Such uses of television proceed together as both artists and audience increasingly become TV-era insiders.

A 1989 magazine essay describes a young man who "sits on one corner of the couch, a 24-inch Sony across the room, a five-inch Hitachi on the table to his right, and a one-and-a-half Panasonic on the coffee table. Sometimes he plays them all at once" (Tennesen). The young man may be the emblem of the immediate future, though multiple screens were foretold when President Lyndon Johnson had three TV monitors installed in the oval office so that he could simultaneously scan each of the three then-major network's broadcast presentations. (Visitors to Graceland, the late Elvis Presley's Memphis, Tennessee, estate, are told that the rock star installed three adjacent televisions in his downstairs bar when he learned about the installation at the Johnson White House.) More recently, the cartoonist Matt Groening draws a cutaway of a house in which several rooms have the favored status of multiple screens, and MTV's videos routinely present multiple screens on-screen, screens within screens. The inset "Smart Window" of the Magnavox Corporation's receiver shows two channels simultaneously, the copy text boasting, "It's like having two TVs in one" (*Rolling Stone*, 564 [Nov. 2, 1989]: 33). Digital television brings a border of screen views from a range of channels, and the split-screen becomes commonplace, the very term, "split-screen life," a referent to late twentieth-century busy-ness.

What about the next generation, who will not only grow up in a house with the television(s) always on, but who will read books, look at comics, see movies, and watch TV about television, even as they sit before screens at work and drive cars with map screens on the dashboard? This is a generation oriented to the screen in videos for infants, a generation for whom hands-on component-part toys like Lego will be less central to childhood than Nintendo and other video games. And it is one that will probably find the interactive on-screen experience carried forward into all sorts of programming formats as, for instance, they select camera angles for sports broadcasts and for music concerts and compete in game shows, all activities currently available in test markets (Pollack).

Technologically imminent is the merging of television with the computer, so that a third TV generation growing up playing Super Mario Brothers at home and taking Computers 101 in school classrooms will find the PC and television merged in the work space. The social analyst, George Gilder, in *Life*

President Lyndon B. Johnson monitoring three network broadcasts simultaneously. Reprinted courtesy of Lyndon Baines Johnson Library.

1990 Magnavox Corporation ad for the dual-screen television.

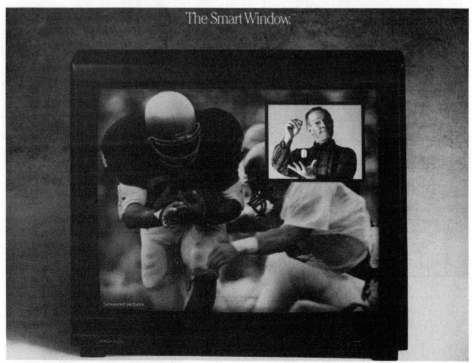

After Television (1990), foresees fiber optics transforming the screen world, and calls this the telecomputer:

> The new system will be the telecomputer, a personal computer adapted for video processing and connected by fiber optic threads to other telecomputers all around the world. Using a two-way system of signals like telephones do, rather than broadcasting one-way like TV, the telecomputer will surpass the television just as the telephone surpassed the telegraph in verbal communication. (19)

A voice like Gilder's takes us back where this discussion began, with American utopian projections of an imminent screen future. The telecomputer, like television, promises to delight and instruct, bringing entertainment and education, giving the individual and family a global reach from the home as viewers—now telekinetic—spend days interacting with celebrity diplomats, religious leaders, film stars, taking "fully interactive" science courses with "the world's most exciting professors" and schooling children at home with "the nation's best teachers imparting one's own cherished moral, cultural, and religious values" (24). The telecomputer, moreover, will allow individuals to "visit" family and friends all over the globe. And "you"—promises the discourse of capitalist individualism—"you could run a global corporation without ever getting on a plane" (24).

Given television's enmeshment with American ideology, it is unsurprising that this discourse presents the new screen era as one to "enhance individualism" and "promote creativity." Instead of the current top-down network and cable "totalitarian" communications structure (one which, since its days are numbered, can now be termed an "alien and corrosive force in democratic capitalism"), the telecomputerized future "will enrich and strengthen democracy and capitalism all around the world" (Gilder, 18–19).

These prognostications ring with the same utopian fervor of the DuMont ads in the 1940s. They construct a public projected both as individualists and families embracing synonymous corporate capitalism and democracy. The telecomputer is both hearth and command post in this discourse, which endorses armchair activism, ignores the epistemic disturbance of the hyperreal, promises to bring the agonizing allegory of the child to a happy conclusion, and to embrace the culture of print, as books, magazines, and other publications are summoned to the screen "edited to your own taste."

This technological utopianism is very familiar from the 1940s. Its resurgence in the early 1990s signals imminent technological change certain to modify the screen environment. As technological change drives social change, with new equipment foreseen and unforeseen, the acculturation of television will inevitably continue. Utopianism aside, the mediative discourses will doubtless continue to disclose angst and aspiration, to participate in ideological arguments in terms both resistant and assenting. But they are certain to be insider discourses, inscribing the continuing social histories of the electronic environment established by television in the second half of this century.

References

Agee, Jonis. *Bend This Heart*. Minneapolis: Coffee House Press, 1989.

Alcott, Louisa May. *Little Women*. 1869, reprint. New York: Macmillan, 1962.

Allen, Robert C., ed. *Channels of Discourse: Television and Contemporary Criticism*. Chapel Hill: University of North Carolina Press, 1987.

——————. *Speaking of Soap Operas*. Chapel Hill: University of North Carolina Press, 1985.

Althusser, Louis. "Ideology and Ideological State Apparatus." In *Video Culture: A Critical Investigation*. John G. Hanhardt, ed. Rochester, New York: Visual Studies Workshop Press, 1986: 56–95.

Anderson, Christopher. "Reflections on *Magnum P.I.*" In *Television: the Critical View*, 4th ed. Horace Newcomb, ed. New York: Oxford University Press, 1987: 112–125.

Anderson, Laurie. "Love Is a Virus." In *Home of the Brave: A Film by Laurie Anderson*. New York and Los Angeles: Warner Brothers, 1986.

Anderson, James A., and Timothy P. Meyer. *Mediated Communication: A Social Action Perspective*. Beverly Hills: Sage, 1988.

Antin, David. "Video: The Distinctive Features of the Medium." In *Video Culture: A Critical Investigation*. John G. Hanhardt, ed. Rochester, New York: Visual Studies Workshop Press, 1986: 147–166.

Arato, Andrew, and Eike Gebhardt, eds. *The Essential Frankfurt School Reader*. New York: Continuum, 1988.

Arlen, Michael J. "What We Do in the Dark." *The Camera Age: Essays on Television*. New York: Farrar Straus Giroux, 1981: 199–204.

——————. "The Air: Prufrock Before the Television Set." *The New Yorker*. Vol. 55 (November 8, 1978): 155–161.

——————. "The Holiday Dinner: A Fable." In *A View from Highway 1*. New York: Farrar, Strauss and Giroux, 1976: 82–89.

Armstrong, W. H. "Mr. Eliot Revisits the Waste Land." *Horizon*. Vol. 5 (January, 1963): 92–93.

Ashbery, John. "Description of a Mask." In *A Wave*. 1984, reprint. New York: Penguin, 1985: 18–30.

Attallah, Paul A. "The Unworthy Discourse: Situation Comedy in Television." *Interpreting Television: Current Research Perspectives.* William D. Rowland, Jr., and Bruce Watkins, eds. Beverly Hills: Sage, 1984: 222–249.

Aurthur, Robert Alan. "Farewell, Farewell, TV." *The Nation.* Vol. 192 (March 4, 1961): 184–187.

Baker, Russell. *The Good Times.* New York: William Morrow, 1989.

—————. "Out of the Suitcase." *The New York Times,* October 4, 1989, p. 23.

Banks, Russell. *Continental Drift.* 1985, reprint. New York: Ballantine, 1986.

—————. *Success Stories.* 1986, reprint. New York: Ballantine, 1987.

Barnouw, Erik. *Tube of Plenty: The Evolution of American Television.* 1975, reprint. New York: Oxford, 1982.

Barthelme, Donald. *Come Back, Dr. Caligari.* New York: Farrar, Straus and Giroux, 1964.

—————. *Guilty Pleasures.* New York: Farrar, Straus and Giroux, 1974.

Barthelme, Frederick. *Second Marriage.* 1984, reprint. New York: Penguin, 1985.

Barthes, Roland. *The Eiffel Tower and Other Mythologies.* Translated by Richard Howard. New York: Hill and Wang, 1979.

—————. *Mythologies.* Translated by Annette Lavers. 1957, reprint. New York: Hill and Wang, 1972.

Batchelor, E. A. "I Was Cured of TV." *Coronet.* Vol. 37 (Feb., 1955): 38–40.

Baudrillard, Jean. *Selected Writings.* Mark Poster, ed. Stanford, California: Stanford University Press, 1989.

Beattie, Ann. *Chilly Scenes of Winter.* 1976, reprint. New York: Warner, 1983.

—————. "Jacklighting." In *The Burning House.* New York: Ballantine, 1983: 17–24.

—————. *Love Always.* 1985, reprint. New York: Vintage, 1986.

Bellah, Robert N., Richard Madsen, William M. Sullivan, Ann Swidler, and Steven M. Tipton. *Habits of the Heart: Individualism and Commitment in American Life.* 1985, reprint. New York: Harper & Row, 1986.

Bellamy, Edward. *Looking Backward—2000–1887.* Reprint. New York: Penguin, 1982.

Bercovitch, Sacvan. *Puritan Origins of the American Self.* New Haven: Yale University Press, 1978.

Berger, John. *Ways of Seeing.* 1972, reprint. New York: Penguin, 1977.

Betsky, Celia. "Inside the Past: The Interior and the Colonial Revival in American Art and Literature." In *The American Colonial Revival.* Alan Axelrod, ed. New York: Norton, 1985: 241–277.

Block, Herbert. *Herblock Through the Looking Glass.* New York: W. W. Norton, 1984.

Bloom, Allan. *The Closing of the American Mind.* New York: Simon and Schuster, 1987.

Boorstin, Daniel J. *The Image: A Guide to Pseudo-Events in America.* 1961, reprint. New York: Atheneum, 1987.

*

Booth, Wayne C. "The Company We Keep: Self-Making in Imaginative Art, Old and New." In *Television: The Critical View*, 4th ed. Horace Newcomb, ed. New York: Oxford University Press, 1987: 382–418.

Boyle, T. Coraghessan. "Heart of a Champion." In *Great Esquire Fiction*. New York: Penguin, 1983: 403–409.

——————. *Greasy Lake and Other Stories*. 1985, reprint. New York: Penguin, 1987.

Bradbury, Ray. *Fahrenheit 451*. 1950, reprint. New York: Ballantine, 1979.

Bradley, David. *No Place to Hide*. 1948. Reprint. Hanover, New Hampshire: University Press of New England, 1983.

Breathed, Berke. *Bloom Country Babylon: Five Years of Basic Naughtiness*. Boston: Little, Brown, 1986.

Byars, Betsy, *The TV Kid*. 1976, reprint. New York: Puffin, 1987.

Cain, William. *F. O. Matthiessen and the Politics of Criticism*. Madison: University of Wisconsin Press, 1988.

Cameron, Peter. *One Way or Another*. New York: Harper & Row, 1986.

Carpenter, Charles H., Jr. "The Tradition of the Old: Colonial Revival Silver for the American Home." In *The Colonial Revival in America*. Alan Axelrod, ed. New York: Norton, 1985: 139–158.

Carver, Raymond. *Cathedral*. 1983, reprint. New York: Vintage, 1984.

Cater, Douglass, and Stephen Strickland. *TV Violence and the Child: The Evolution and Fate of the Surgeon General's Report*. New York: Russell Sage, 1975.

Caufield, Catherine. *Multiple Exposures: Chronicles of the Radiation Age*. Chicago: University of Chicago Press, 1989.

Cerf, Bennett. "Books." *Saturday Review of Literature*. Vol. 31 (June 5, 1948): 4–6.

Chase, Richard. *The American Novel and Its Tradition*. Garden City, New York: Doubleday, 1957.

Connell, Evan S. *Mrs. Bridge*. 1958, reprint. Berkeley: North Point Press, 1981.

Coover, Robert. *Pricksongs and Descants*. 1969, reprint. New York: Penguin, 1970.

Corn, Joseph, ed. *Imagining Tomorrow: History, Technology, and the American Future*. Cambridge: MIT Press, 1986.

Cronkite, Walter. "Television and the News." In *Twenty-Three Views of Television*. New York: Holt, Rinehart and Winston, 1962: 227–240.

Crosby, John. "What's TV Going to Do to Your Life?" *House Beautiful*. Vol. 92 (Feb., 1950): 67, 87–88.

Cross, Donna Woolfolk. *Media-Speak: How Television Makes Up Your Mind*. 1983, reprint. New York: New American Library, 1984.

Cutright, Melitta. *The National PTA Talks to Parents: How to Get the Best Education for Your Child*. New York: Doubleday, 1989.

Czitrom, Daniel J. *Media and the American Mind: From Morse to McLuhan*. Chapel Hill: University of North Carolina Press, 1982.

"Dark (Screen) Future." *Time*. Vol. 58 (July 30, 1951): 53.

Davidson, Cathy N., ed. *Reading in America: Literature and Social History*. Baltimore: Johns Hopkins University Press, 1989.

Davis, Erik. "TV's Fascinating, Frightening Future." *The Utne Reader*. No. 40 (July/Aug., 1990): 86–87.

De Certeau, Michel. *The Practice of Everyday Life*. Trans. Steven Rendall. 1984, reprint. Berkeley, Los Angeles: University of California Press, 1988.

De George, Richard T. *The Nature and Limits of Authority*. Lawrence, Kansas: University Press of Kansas, 1985.

De Grazia, Sebastian. *Of Time, Work, and Leisure*. New York: The Twentieth Century Fund, 1962.

DeLillo, Don. *Libra*. New York: Viking, 1988.

—————. *Players*. 1977, reprint. New York: Vintage, 1984.

—————. *White Noise*. 1985, reprint. New York: Penguin, 1986.

Diamond, Edwin. *Good News, Bad News*. Cambridge: MIT Press, 1978.

—————. *Sign Off: The Last Days of Television*. Cambridge: MIT Press, 1982.

—————. *The Tin Kazoo: Television, Politics, and the News*. Cambridge, 1975.

"Diary of a Viewer." *The New Yorker*. Vol. 23 (August 30, 1947): 44, 46–48, 50–55.

Dicken-Garcia, Hazel. *Journalistic Standards in Nineteenth-Century America*. Madison: University of Wisconsin Press, 1989.

Dimock, Wai-Chi. *Empire for Liberty*. Princeton: Princeton University Press, 1988.

Eco, Umberto. *Travels in Hyperreality*. 1983, reprint. San Diego: Harcourt Brace Jovanovich, 1986.

"Editorial." *Special Reports: The Family*. Aug.–Sept., 1989: 3.

Elkin, Stanley. *Stanley Elkin's The Magic Kingdom*. 1985, reprint. New York: Dutton, 1986.

Elkind, David. *The Hurried Child: Growing Up Too Fast Too Soon*. Reading, Massachusetts: Addison-Wesley, 1988.

Ellis, Bret Easton. *Less Than Zero*. 1985, reprint. New York: Penguin, 1985.

Ellis, Trey. *Platitudes*. New York: Vintage, 1988.

Emerson, R. W., secundus. "Television's Peril to Culture." *The American Scholar*. Vol. 19, No. 2 (Spring, 1950): 137–140.

Emerson, Ralph Waldo. "Self-Reliance." In *Selections from Ralph Waldo Emerson*. Ed. Stephen Whicher. Boston: Houghton, Mifflin, 1960.

Enzensberger, Hans M. "Constituents of a Theory of the Media." *Video Culture: A Critical Investigation*. John G. Hanhardt, ed. Rochester, New York: Visual Studies Workshop Press, 1986: 96–123.

Ephron, Nora. *Heartburn*. 1983, reprint. New York: Pocket Books, 1986.

Evans, Rowland, Jr. "A New Tool for Politics." In *The Eighth Art: Twenty Three Views of Television*. New York: Holt, Rinehart and Winston, 1962: 39–53.

"Family Recreation." *Recreation*. Vol. 46 (Nov., 1952): 340.

Faught, Millard C. "Television: What Potentialities! What Progress! What Problems!" *Vital Speeches*. Vol. 17 (April 15, 1951): 413–416.

Feuerlicht, Ignace. *Alienation: From the Past to the Future*. Westport, Conn.: Greenwood Press, 1978.

Fiske, John. *Television Culture*. London, New York: Methuen, 1987.

Fitzgerald, F. Scott. *The Great Gatsby*. 1925, reprint. New York: Scribner's, 1953.

————. *The Last Tycoon*. 1941, reprint. New York: Scribner's, 1988.

Floyd, Vircher B. "Creative Leisure." *Recreation*. Vol. 51 (Oct., 1958): 269.

Ford, Richard. *The Sportswriter*. New York: Vintage, 1986.

Fried, Michael. *Realism, Writing, Disfiguration: on Thomas Eakins and Stephen Crane*. Chicago: University of Chicago Press, 1987.

Friedan, Betty. *The Feminine Mystique*. 1963, reprint. New York: Dell, 1983.

Gans, Herbert J. *Middle American Individualism: The Future of Liberal Democracy*. New York: Free Press, 1988.

Gardner, John. *October Light*. New York: Knopf, 1977.

Gerbner, George, and Larry Gross, Michael Morgan, Nancy Signorielli. "Charting the Mainstream: Television's Contribution to Political Orientations." *American Media and Mass Culture: Left Perspectives*. Edited by Donald Lazere. Berkeley and Los Angeles: University of California Press, 1987: 441–464.

Gerbner, George, and Larry Gross. "The Scary World of TV's Heavy Viewer." *Psychology Today*. Vol. 9 (April, 1976): 41–45, 89.

Gilbert, Sandra, and Susan Gubar. "Masterpiece Theatre: An Academic Melodrama." Lecture Presented June 26, 1990 at The Wyoming Conference on English, Laramie, Wyoming.

Gilder, George. *Life After Television: The Coming Transformation of Media and American Life*. Knoxville: Whittle, 1990.

Gitlin, Todd. *The Sixties: Years of Hope, Days of Rage*. New York: Bantam, 1987.

————, ed. *Watching Television*. New York: Pantheon, 1986.

————. *The Whole World Is Watching: Mass Media in the Making of the New Left*. Berkeley, Los Angeles: University of California Press, 1980.

Glasser, William, M.D. *Schools without Failure*. 1969, reprint. New York: Harper & Row, 1975.

Goodis, David. *Street of No Return*. New York, 1954.

Goodman, Paul. "Leisure: Purposeful or Purposeless." 1959, reprint. *Recreation in America*. Pauline Madow, ed. New York: W. H. Wilson, 1951: 30–31.

Gottlieb, Annie. *Do You Believe in Magic: Bringing the Sixties Home*. 1987, reprint. New York: Simon and Schuster, 1988.

"Grandmother's Houses." *The Household*. 5. 1872: 115–119.

Guthrie, Tyrone. "Theatre and Television." In *The Eighth Art: Twenty Three Views of Television*. New York: Holt, Rinehart and Winston, 1962: 91–99.

Hall, Trish. "Writing Like MTV: The Making of a Cult." *The New York Times*. June 11, 1990: 84.

Halprin, Jeffrey A. "Getting Back to Work: The Revaluation of Work in American

Literature and Social Theory, 1950–1985." Ph.D. Dissertation. Boston University, 1987.

Hartley, John. "Encouraging Signs: Television and the Power of Dirt, Speech, and Scandalous Categories." *Interpreting Television: Current Research Perspectives*. Willard D. Rowland, Jr., and Bruce Watkins, eds. Beverly Hills: Sage, 1984, 119–141.

Haskell, Thomas L., ed. *The Authority of Experts: Studies in History and Theory*. Bloomington: Indiana University Press, 1984.

Hawthorne, Nathaniel. *The Scarlet Letter*. 1850, reprint. New York: Penguin, 1983.

————. *Selected Tales and Sketches*. Hyatt Waggoner, ed. New York: Holt, Rinehart and Winston, 1970.

Heckscher, August. "Foreword." In *Of Time, Work, and Leisure*. Sebastian de Grazia, ed. New York: The Twentieth Century Fund, 1962.

Henry, William A. III. "The Meaning of TV." *Life*. March, 1989: 66–75.

Hersh, Seymour M. *The Price of Power: Kissinger in the Nixon White House*. New York: Summit, 1983.

Heuman, William. "Her Shadow Love." *The Saturday Evening Post*. Vol. 227 (April 12, 1955): 29, 85, 88–98.

Hijuelos, Oscar. *Mambo Kings Sing Songs of Love*. New York: Farrar, Straus and Giroux, 1989.

Hoffman, Daniel. *Form and Fable in American Fiction*. 1961, reprint. New York: Oxford University Press, 1965.

Honey, Maureen. *Creating Rosie the Riveter: Class, Gender, and Propaganda During World War II*. Amherst: University of Massachusetts Press, 1984.

Hornaday, Mary. "End of Conversation?" *The Christian Science Monitor*. (March 19, 1949): 5.

Howard, Leon. *Literature and the American Tradition*. Garden City, New York: Doubleday, 1960.

Hughes, Robert. "Giving Success a Good Name." *Time*. June 20, 1988: 76–79.

————. "The Patron Saint of Neo-Pop." *The New York Review of Books*, XXXVI, No. 9 (June 1, 1989): 29–32.

Humanities in America: A Report to the President, the Congress, and the American People. Lynne V. Cheney, Chairman. Washington, D.C.: National Endowment for the Humanities, 1988.

Huston, Paul G. *Around an Old Homestead: A Book of Memories*. Cincinnati, 1906.

"Hypnosis in Your Living Room." *The Architectural Forum*. Sept., 1948, reprint. *Reader's Digest*. Vol. 54 (April, 1949): 70–72.

James, Caryn. "Characters from the Screen Are Invading the Printed Page." *The New York Times*. May 31, 1990: C15, C19.

Jarrell, Randall. "A Sad Day at the Supermarket." 1960, reprint. *Recreation in America*. Pauline Madow, ed. New York: W. H. Wilson, 1965: 157–167.

Jehlen, Myra. *American Incarnation*. Cambridge: Harvard University Press, 1986.

Johns, Elizabeth. *Thomas Eakins: The Heroism of Modern Life*. Princeton: Princeton University Press, 1983.

Josephson, Matthew. *Portrait of the Artist as American*. 1930, reprint. New York: Octagon, 1964.

Joyrich, Lynne. "All that Heaven Allows: TV Melodrama, Postmodernism and Consumer Culture." *Camera Obscura: A Journal of Feminism and Film Theory*. No. 16 (January, 1988): 128–153.

Kaiser, Charles. *1968 in America: Music, Politics, Chaos, Counterculture, and the Shaping of a Generation*. New York: Weidenfield and Nicholson, 1988.

Kaplan, E. Ann. *Rocking Around the Clock*. New York: Methuen, 1987.

Kaplan, Max. *Leisure in America: A Social Inquiry*. New York: John Wiley & Sons, 1960.

Kaplan, Morton A. *Alienation and Identification*. New York: Free Press, 1976.

Keating, Susan Katz. "Much Ink Spilled on Minimal Topic." *Insight*. February 13, 1989: 60–62.

Kernan, Alvin. *The Death of Literature*. New Haven: Yale University Press, 1990.

Kerouac, Jack. *The Dharma Bums*. 1958, reprint. New York: New American Library, 1959.

Kosinski, Jerzy. *Being There*. New York: Harcourt Brace Jovanovich, 1970.

Kozinn, Allan. "Fun with Jack Benny and Electronic Music." *The New York Times* (National Edition). Sept. 27, 1989, p. 17.

Kraus, Richard. *Recreation Today*. New York: Appleton, Century, Crofts, 1966.

Lazere, Donald, ed. *American Media and Mass Culture: Left Perspectives*. Berleley, Los Angeles: University of California Press, 1987.

Leach, Penelope. *Your Growing Child: From Babyhood through Adolescence*. 1983, reprint. New York: Knopf, 1989.

Leavitt, David. *Family Dancing*. New York: Knopf, 1985.

Lee, Robert. *Religion and Leisure in America: A Study in Four Dimensions*. New York, Nashville: Abingdon, 1964.

Leithauser, Brad. "Any Place You Want." *The New York Review of Books*. Vol. 37, No. 4 (March 15, 1990): 7–10.

Lesser, Gerald. *Children and Television: Lessons from Sesame Street*. New York: Random House, 1974.

Lewis, R. W. B. *The American Adam: Innocense, Tragedy, and Tradition in the Nineteenth Century*. University of Chicago Press, 1955.

Leyner, Mark. *My Cousin, My Gastroenterologist*. New York: Harmony, 1990.

Lieberson, Jonathan. "TV: A Day in the Life." *The New York Review of Books*. Vol. 36, No. 6 (April 13, 1989): 15–16, 18–20.

Longfellow, Henry Wadsworth. *The Poetical Works*. Boston: Houghton, Mifflin and Co., 1883.

Lora, Ronald, ed. *America in the '60s: Cultural Authorities in Transition*. New York: John Wiley & Sons, 1974.

Lukacs, John. *A New History of the Cold War*. Garden City, New York: Doubleday, 1966.

Macaulay, David. *Motel of the Mysteries*. Boston: Houghton Mifflin, 1979.

Magli, Patrizia. "The Face and the Soul." *Fragments for a History of the Human Body*. Part Two. Michel Feher, ed. New York: Zone, 1989: 87–127.

Mailer, Norman. *An American Dream*. 1964, reprint. New York: Dell, 1965.

"Making Peace with the Tube." *Special Reports: The Family*. Aug.–Sept., 1989: 5.

Manchester, Harland. "TV Will Change You." *Nation's Business*. Vol. 37 (June, 1949): 40–42, 74.

Mander, Jerry. *Four Arguments for the Elimination of Television*. 1977, reprint. New York: Quill, 1978.

Mann, Denise. "The Spectacularization of Everyday Life: Recycling Hollywood Stars and Fans in Early Television Variety Shows." *Camera Obscura: A Journal of Feminism and Film Theory/* 16. January, 1988: 49–77.

Mannes, Marya. "The Lost Tribe of Television." In *The Eighth Art: Twenty Three Views of Television Today*. New York: Holt, Rinehart and Winston, 1962: 23–29.

Marc, David. *Demographic Vistas: Television in American Culture*. Philadelphia: University of Pennsylvania Press, 1984.

Marcuse, Herbert. *One-Dimensional Man*. Boston: Beacon, 1964.

Maritain, Jacques. "Work and Leisure." In *Recreation on America*. Pauline Madow, ed. New York: H. W. Wilson, 1965: 28–31.

Markfield, Wallace. "Oh, Mass Man! Oh, Lumpen Lug! Why Do You Watch TV?" *The Saturday Evening Post*. Vol. 241 (Nov. 30, 1968): 28–29, 72.

Markoff, John. "Personal Computers May Soon Add TV's Beauty." *The New York Times*. National Edition. Sept. 12, 1989, pp. 1, 33.

Marling, Karal Ann. *George Washington Slept Here: Colonial Revivals and American Culture, 1876–1986*. Cambridge: Harvard University Press, 1988.

Marschall, Rick. *The Golden Age of Television*. New York: Bison Books, 1987.

Mason, Bobbie Ann. *Shiloh and Other Stories*. 1982, reprint. New York: Harper & Row, 1985.

——————. *In Country*. 1985, reprint. New York: Harper & Row, 1986.

——————. *Love Life*. New York: Harper & Row, 1989.

Matthiessen, F. O. *American Renaissance: Art and Expression in the Age of Emerson and Whitman*. 1941, reprint. New York: Oxford University Press, 1968.

May, Elaine Tyler. *Homeward Bound: American Families in the Cold War Era*. New York: Basic Books, 1988.

McCorkle, Jill. *The Cheer Leader*. 1984, reprint. New York: Penguin, 1985.

McCrumb, Sharyn. *If Ever I Return, Pretty Peggy-O*. New York: Scribner's, 1990.

McLuhan, Marshall. *Understanding Media: The Extensions of Man*. New York: NAL Penguin, 1964.

McMillan, Terry. *Disappearing Acts*. New York: Viking, 1989.

Melville, Herman. *Moby Dick*. 1851, reprint. New York: W. W. Norton, 1967.

Mencken, H[enry] L[ouis]. "Postscripts to the American Language: Video Verbiage." *The New Yorker.* Vol. 24 (Dec. 11, 1948): 112–116.

Meyrowitz, Joshua. *No Sense of Place: The Impact of Electronic Media on Social Behavior.* New York: Oxford University Press, 1985.

Miller, Mark Crispin. *Boxed-In: The Culture of TV.* Evanston, Illinois: Northwestern University Press, 1988.

Mills, C. Wright. *White Collar: The American Middle Classes.* 1951, reprint. New York: Oxford University Press, 1977.

Milward, John. "Big Is Better." *Special Reports: The Family.* Aug.–Oct., 1989: 65.

Minot, Susan. *Monkeys.* New York: E. P. Dutton/ Seymour Lawrence, 1986.

Montagu, Ashley. "Television and the New Image of Man." In *The Eighth Art: Twenty-Three Views of Television Today.* Robert Lewis Shayon, ed. New York: Holt, Rinehart and Winston, 1962: 125–134.

Morley, David. *Family Television: Cultural Power and Domestic Leisure.* London: Comedia, 1986.

Mungo, Raymond. *Famous Long Ago.* Boston: Beacon Press, 1970.

Mulvey, Laura. *Visual and Other Pleasures.* Bloomington: Indiana University Press, 1989.

Nesbit, Robert. *The Twilight of Authority.* New York: Oxford University Press, 1975.

"The New Organization Man." *U.S. News & World Report.* January 16, 1989: 47–49.

Noonan, Peggy. *What I Saw at the Revolution: A Political Life in the Reagan Era.* New York: Random House, 1990.

Olmstead, Robert. *Soft Water.* New York: Vintage, 1988.

Orvell, Miles. *The Real Thing: Imitation and Authenticity in American Culture.* Chapel Hill: University of North Carolina Press, 1989.

Palmer, Patricia. *The Lively Audience: A Study of Children Around the TV Set.* Sydney, London, Boston: Allen and Unwin, 1986.

Pease, Donald. *Visionary Compacts: American Renaissance Writings in Cultural Context.* Madison: University of Wisconsin Press, 1987.

Percy, Walker. *Lancelot.* 1977, reprint. New York: Avon, 1978.

——————. *Lost in the Cosmos: The Last Self-Help Book.* New York: Farrar, Straus & Giroux, 1983.

——————. *The Moviegoer.* 1960, reprint. New York: Avon, 1982.

Pittman, Robert W. "We're Talking the Wrong Language to 'TV Babies.'" *The New York Times.* January 24, 1990: A15.

Pollack, Andrew. "New Interactive TV Threatens the Bliss of Couch Potatoes." *New York Times.* June 18, 1990: A1, C7.

Pope-Hennessy, John. *The Portrait in the Renaissance.* New York: Pantheon, 1966.

Postman, Neil. *Amusing Ourselves to Death: Public Discourse in the Age of Show Business.* 1985, reprint. New York: Penguin, 1986.

Price, Reynolds. *Real Copies*. North Carolina: North Carolina Wesleyan College Press, 1988.

Pulos, Arthur J. *The American Design Adventure 1940–1975*. Cambridge: MIT Press, 1988.

Pynchon, Thomas. *Vineland*. Boston: Little, Brown, 1990.

Reynolds, Sir Joshua. *Discourses on Art*. Chicago: Packard, 1945.

Reynolds, Ralph. "So You Can't Kick the TV Habit." *The Saturday Evening Post*. Vol. 248 (Sept., 1976): 24.

Rhoads, William B. *The Colonial Revival*, 2 Vols. New York: Garland, 1977.

——————. "The Colonial Revival and the Americanization of Immigrants." In *The Colonial Revival in America*. Alan Axelrod, ed. New York: Norton, 1985: 341–361.

Riesman, David. *Individualism Reconsidered*. Glencoe, Illinois: The Free Press, 1953.

——————. With Nathan Glazer and Reul Denny. *The Lonely Crowd*. 1950, reprint. New Haven: Yale University Press, 1969.

Ritts, Paul. *The TV Jeebies*. Illustrated by Dick Strome. Philadelphia, Toronto: John C. Winston, 1951.

Robinson, Gertrude Joch. "Television News and the Claim to Facticity." In *Interpreting Television: Current Research Perspectives*. Willard D. Rowland, Jr., and Bruce Watkins, ed. Beverly Hills: Sage, 1984, 199–221.

Rosen, Norma. *Touching Evil*. New York: Harcourt, Brace and World, 1969.

Ross, Andrew. *No Respect: Intellectuals and Popular Culture*. New York: Routledge, 1989.

Rosten, Leo. "What Television Can—and Cannot—Do." *Reader's Digest*. Vol. 82 (Feb., 1963): 143–144, 146.

Rudnick, Paul. *I'll Take It*. New York: Knopf, 1989.

"Salute to Television." *Life*. Vol. 25, No. 23 (Dec. 6, 1948): 50.

Scarpelli, Henry. *TV TEE-HEES*. New York: Fleet, 1963.

Schmitt, Jean-Claude. "The Ethics of Gesture." In *Fragments for a History of the Human Body*, Part Two. Michel Feher, ed. New York: Zone, 1989: 129–147.

Schorr, Daniel. "Is There Life After TV?" *Esquire*. Vol. 88 (Oct., 1977): 105–106, 156–164.

Schwoch, James. "Presenting the History of the Future: Television Advertising and Telecommunications Discourse in American Culture." Milwaukee: University of Wisconsin-Milwaukee Center for Twentieth Century Studies. Working Paper No. 11 (Fall, 1988): 1–16.

"Sedulus." "Pickling Our Minds." *The New Republic*. Vol. 168 (June 23, 1973): 25.

"Sedulus." "Television: Seeing Isn't Believing." *The New Republic*. Vol. 164 (March 13, 1971): 30–32.

Segal, Howard. *Technological Utopianism in American Culture*. Chicago: University of Chicago Press, 1985.

Seldes, Gilbert. "The 'Errors' of Television." *The Atlantic Monthly.* Vol. 159 (May, 1937): 531–541.

Sennett, Richard. *Authority.* New York: Knopf, 1980.

Simpson, Mona. *Anywhere But Here.* 1986, reprint. New York: Vintage, 1988.

Sissman, L. E. "Innocent Bystander: Facing the Tube." *Atlantic Monthly.* Vol. 233 (Feb., 1974): 25–27.

Smith, David R. *Masks of Wedlock: Seventeenth-Century Dutch Marriage Portraiture.* Ann Arbor: UMI Research Press, 1982.

Smith, Lee. *Black Mountain Breakdown.* New York: Putnam's, 1980.

Spigel, Lynn. "Installing the Television Set: Popular Discourses on Television and Domestic Space, 1948–1955." *Camera Obscura: A Journal of Feminism and Film Theory.* No. 16 (Jan., 1988): 11–46.

Spock, Benjamin. *Dr. Spock on Parenting.* New York: Simon and Schuster, 1988.

Spock, Benjamin, M.D. *Dr. Spock on Parenting.* 1988, reprint. New York: Pocket Books, 1989.

Steinberg, Saul. *The Inspector.* New York: Viking, 1973.

Steiner, Wendy. *Exact Resemblance to Exact Resemblance: The Literary Portraiture of Gertrude Stein.* New Haven, London: Yale University Press, 1978.

Stich, Sidra. *Made in U S A: An Americanization in Modern Art, The '50s & '60s.* Berkeley, Los Angeles: The University of California Press, 1987.

Stockman, David A. *The Triumph of Politics: The Inside Story of the Reagan Revolution.* 1986, reprint. New York: Avon, 1987.

Stone, Elizabeth. "Our New Neighborhood." *Special Reports: The Family.* Aug.-Sept., 1989: 9.

Stowe, Harriet Beecher. *House and Home Papers.* Boston, 1865.

——————. *Uncle Tom's Cabin.* 1852, reprint. New York: Oxford University Press, 1965.

Television and Behavior: Ten Years of Scientific Progress and Implications for the Eighties, Volume I: Summary Report; Volume II: Technical Reviews. Rockville, Md.: U.S. Department of Health and Human Services, 1982.

Television and Growing Up: The Impact of Televised Violence; Report to the Surgeon General, United States Public Health Service from the Surgeon General's Scientific Advisory Committee on Television and Social Behavior. Rockville, Maryland: National Institute of Mental Health, 1972.

Tennesen, Michael. "The TV Addict." *Special Report: On Family.* Aug.–Oct., 1989: 65.

Tichi, Cecelia. "Television and Recent Fiction." *American Literary History.* Vol. 1, No. 1 (Spring, 1989): 110–130.

Trudeau, G[arry] B. *Bravo for Life's Little Ironies* New York: Popular Library, 1973.

——————. *Call Me When You Find America.* 1973, reprint. New York: Bantam, 1976.

——————. *Guilty, Guilty, Guilty!* 1974, reprint. New York: Bantam, 1976.

—————. *Dare to Be Great, Ms. Caucus.* 1975, reprint. New York: Bantam, 1976.

—————. *Wouldn't a Gremlin Have Been More Sensible?* 1975, reprint. New York: Bantam, 1976.

—————. *The President Is a Lot Smarter Than You Think.* New York: Popular Library, 1973.

—————. *I Have No Son.* New York: Popular Library, 1973.

—————. *And That's My Final Offer.* New York: Holt, Rinehart and Winston, 1980.

—————. *You Give Great Meeting, Sid!* New York: Holt, Rinehart and Winston, 1983.

—————. *Downtown Doonesbury.* New York: Holt, Rinehart and Winston, 1987.

—————. *The People's Doonesbury: Notes from Underfoot, 1978–1980.* New York: Holt, Rinehart and Winston, 1981.

—————. *He's Never Heard of You, Either.* New York: Holt, Rinehart and Winston, 1981.

Turow, Scott. *Presumed Innocent.* 1987, reprint. New York: Warner, 1988.

TV Guide Roundup. The Editors of *TV Guide.* New York: Holt, Rinehart and Winston, 1960.

"TV and the Recreation Program." *Recreation.* Vol. 46 (June, 1953): 189.

"TV: An Interim Summing-Up." *The Saturday Review of Literature.* Vol. 33 (Aug. 26, 1950): 7–8, 29–34.

"TV the Terrible." *Fortune.* Vol. 42 (July, 1950): 55–58.

Updike, John. *Rabbit Redux.* 1971, reprint. New York: Fawcett, 1972.

—————. *Roger's Version.* 1986, reprint. New York: Ballantine, 1987.

Utley, Clifton M. "How Illiterate Can Television Make Us?" *The Commonweal.* Vol. 49 (Nov. 19, 1948): 137–139.

Vigarello, Georges. "The Upward Training of the Body from the Age of Chivalry to Courtly Civility." *Fragments for a History of the Human Body,* Part Two. Michel Feher, ed. New York: Zone, 1989: 149–199.

Wallace, David Foster. "Fictional Futures and the Conspicuously Young." *Review of Contemporary Fiction.* Vol. 8, No. 3 (Fall, 1988): 36–49.

—————. *Girl with Curious Hair.* New York: Norton, 1989.

Ward, John William. *Red, White, and Blue: Men, Books, and Ideas in American Culture.* New York: Oxford University Press, 1969.

Warner, William. "August 29, 1982 Bobst Library." *ISBN.* New York: Farrah, Upland, Westmoreland, and Granger, 1983.

"What TV Is Doing to America." *U.S. News & World Report.* Vol. 39 (Sept. 2, 1955): 36–76.

"What TV Is Doing to Home Life." *U.S. News & World Report.* Vol. 39 (Sept. 2, 1955): 45–48.

Wheeler, Judith. "The Electronic Age." *The Saturday Review.* Vol. 49 (June 2, 1966): 21–22.

Whitbread, Jane, and Vivian Cadden. "The Real Menace of TV." *Harper's Magazine*. Vol. 209 (Oct., 1954): 81–83.

White, Mimi. "Ideological Analysis and Television." In *Channels of Discourse: Television and Contemporary Criticism*. Robert C. Allen, ed. Chapel Hill: University of North Carolina Press, 1987: 134–171.

Whithey, Stephen, and Ronald P. Abeles, eds. *Television and Social Behavior: Beyond Violence and the Child*. Hillsdale, N.J.: Lawrence Erlbaum, 1980.

Whittemore, Reed. "The Big Picture." *The New Republic*. Vol. 174 (February 7, 1976): 21–23.

Whittier, John Greenleaf. *Snow-Bound. The Harper American Literature*, Vol. II. Donald McQuade et al., eds. New York: Harper & Row, 1986.

Wiener, Norbert. *The Human Use of Human Beings: Cybernetics and Society*. New York: Putnam, 1950.

Williams, Raymond. *Television: Technology and Cultural Form*. 1974, reprint. New York: Schocken, 1975.

Willingham, Calder. "Television: Giant in the Living Room." *American Mercury*. Vol. 74 (Feb., 1952): 114–119.

Wilson, Sloan. *The Man in the Gray Flannel Suit*. 1955, reprint. New York: Simon and Schuster, 1979.

Wilson, T. W., Jr. "TV and the Recreation Department." *Recreation*. Vol. 47 (May, 1954): 290–292.

Winn, Marie. "The Plug-In Drug." *The Saturday Evening Post*. Vol. 249 (Nov., 1977): 40–41, 90–91.

————. *The Plug-In Drug*. New York: Viking, 1977.

Wolitzer, Meg. *This Is Your Life*. New York: Crown, 1988.

Wright, Frank Lloyd. *An Autobiography*. 1932, reprint. New York: Horizon, 1977.

Wright, Lawrence. *In the New World: Growing Up with America 1960–1984*. New York: Knopf, 1988.

Yoder, Robert M. "Be Good! Television's Watching." *The Saturday Evening Post*. Vol. 221 (May 14, 1949): 29, 131–133.

York, Max. "We've Come a Long Way, Nintendo-ly." *The Tennessean*. July 22, 1989: 5-D.

Zuboff, Shoshana. *In the Age of the Smart Machine: The Future of Work and Power*. New York: Basic Books, 1988.

Index